BASIC mechanical vibrations

Butterworths BASIC Series includes the following titles:

BASIC aerodynamics
BASIC hydraulics
BASIC hydrology
BASIC materials studies
BASIC matrix methods
BASIC mechanical vibrations
BASIC numerical mathematics
BASIC soil mechanics
BASIC statistics
BASIC stress analysis
BASIC thermodynamics and heat transfer

BASIC mechanical vibrations

A J Pretlove, BSc(Eng), FIMechE, FIoA, CEng
Department of Engineering, University of Reading

Butterworths
London Boston Durban Singapore Sydney Toronto Wellington

First published 1985

© Butterworth & Co. (Publishers) Ltd 1985

British Library Cataloguing in Publication Data

Pretlove, A. J.
  BASIC mechanical vibrations.
  1. Vibration
  I. Title
  531′.32   TA355
  ISBN 0–408–01554–3

Library of Congress Cataloguing in Publication Data

Pretlove, A. J.
  BASIC mechanical vibrations.
  Bibliography: p.
  Includes index.
  1. Vibration. 2. Vibration—Computer programs.
  3. Basic (Computer program language) I. Title.
  TA355.P72 1985   620.3′028′5424   84–23121
  ISBN 0–408–01554–3

Typeset by Mid-County Press, 2a Merivale Road, London
Printed and bound in Great Britain by Anchor Brendon Ltd, Tiptree, Essex

Preface

Vibrations is a subject of great importance in many fields of engineering design. It is, perhaps, most important in mechanical, marine and aeronautical engineering where vibration can cause wear in bearings, structural fatigue in load-bearing components, and other ill-effects. It is also becoming increasingly important in civil engineering as a result of more efficient design of high strength reinforced and prestressed concrete structures. This process has led to relatively slender structures, particularly bridges and tall buildings, which are sufficiently flexible for vibration to be a serious irritant to human beings using the structure. There is also increasing anxiety that fatigue failure may occur in such structures. Thus, design to counteract the ill-effects of vibration is a topic of increasing importance. Vibrations can also be used to perform useful functions such as, for example, in concrete compactors, ultrasonic cleaning tanks and dynamic vibration absorbers (see Chapter 4). Thus the design engineer needs to know (1) how to calculate natural frequencies of vibration, (2) how to calculate amplitudes of forced motion, principally in terms of deflection or stress, (3) how to attenuate unwanted vibrations, (4) how to measure vibrations, and (5) how to interpret these measurements.

This book, therefore, is devoted to a presentation of vibrations which combines basic theory with the development of useful computer programs to make design calculations. Vibration theory uses a wide range of mathematics and some of the programs given here are intended for rather general use such as, for example, Program 3.1 for solving quadratic equations. Other programs are rather more specific to design.

Much of the analysis of engineering design is today couched in terms of matrix algebra and arithmetic. This is because computers are now used to solve such problems and a formalization of calculation is essential for efficient and orderly calculation of results. The programs in the book are written in BASIC and should therefore be of direct use to most readers through access to a microcomputer or time-sharing system. Although some such systems have a set of matrix algebra instructions as hardware it is probable that most do not. As a result some subroutines have been included for performing simple matrix operations on two-dimensional arrays which can be used in vibration calculations.

Many of the figures in the book consist of conventionally represented sets of mass and spring elements. These, of course, are idealized models of real systems. One of the most difficult problems facing the design engineer is the assembly of such a model for his real system which, as far as vibration is concerned, accurately represents its behaviour. In several places in the book an indication is given of how this modelling can be achieved.

Limitations on space and the wish to make the book basic (as well as BASIC) has meant that the coverage is restricted to vibrations of linear systems. Most real systems contain non-linear components, particularly the damping element, but, paradoxically, most real problems are solved quite accurately by means of a linear treatment. Thus, no apology is really needed for the treatment given here except to say that the reader should be aware that there are some problems in real life where non-linear analysis is required. The other important matter not treated here is the subject of random vibrations. This topic is relatively advanced (mathematically) and computations concerned with it are usually quite elaborate. Thus, while it is a matter of great importance to the designer, particularly where fatigue calculations are concerned, it is inappropriate here.

This book should be useful to those engineers who need to make vibration design calculations. It also contains some programs in BASIC which, while being useful for purposes connected with vibrational design, will also be useful more generally in the design process. It is hoped therefore that the book will have a wide appeal. Much of the material for the book is derived from material used for teaching vibrations to undergraduates, both at Reading and for an intensive computer-based course given, by invitation, at the University of Malta. Thus, it is also a useful university text for students of mechanical engineering.

The author is grateful to his colleagues John Turner at Reading and Reto Cantieni of the Swiss Federal Materials Testing Laboratory (EMPA) at Dubendorf, both of whom have offered constructive criticism of the text: also to Jan Stevens who prepared the diagrams. The text was typed and corrected by the author on that marvellous thing, the word-processor program on his microcomputer.

AJP
1985

Contents

Chapter 1

The BASIC computer language

This book forms one of a series dedicated to practical computing and it has been written to teach the use of the BASIC computer language in solving practical vibration problems. This chapter is a brief introduction to computing with special emphasis on fundamentals of BASIC. The later chapters give concise elements of vibration theory followed by problem solving examples making use of BASIC programs. A plentiful supply of both worked and unworked examples will give the student 'hands-on' teaching of computing together with that special insight into theory which only a logical programming of examples can give.

The early parts of this chapter are dedicated especially to the microcomputer user. Most microcomputers are delivered with only the briefest account of how they are structured and how they work. The user is generally expected to regard the machine as a 'black box' and to ask no questions about how BASIC operates. This may be a satisfactory starting point but the user may well want to know how it is done. The first half of this chapter is therefore devoted to this task. The second half gives an account of the BASIC language as used in the rest of the book. Further reading at the end of the book is intended for those who wish to extend their knowledge of computers, BASIC and vibrations beyond the content of this book.

1.1 Computer structure

A computer is little more than an adding machine with storage space for numbers and the ability to follow a set of operations in sequence, called a program. The heart of a computer is the so-called Central Processing Unit (CPU for short) which in modern machines consists of a single silicon chip, the microprocessor chip. It is here that all the work is done; not only the calculations, but also the sequential processing of the program. In the early days of computers this part of the machine was called the 'mill', and this is an apt description. In following the program the CPU will collect the next instruction from the store, interpret it so that it knows what is required, collect relevant numbers from another part of the store, conduct the calculation and return the result to store. It then moves on to the next instruction. This single cycle of operation can be very rapid, taking about one millionth of a second to perform in many machines. Thus quite complicated calculations can be performed by means of a sequence of very simple instructions and in a short time. At this level it is necessary to make instructions simple, rapid and efficient, and so a special and rather obscure language called 'machine code' is used. This language is specific to the microprocessor chip being used. Thus, for example, the 6502 microprocessor used in an Apple or PET computer will have a machine code different from the Z80 microprocessor used in a Sinclair Spectrum.

It is essential to realize that the microprocessor can only work with a program in its own code and so all programs in other languages (including BASIC) have to be translated within the machine. The machine code is therefore called a low-level language whereas BASIC and other languages which have to be translated, such as FORTRAN, PASCAL and ALGOL, are called high-level languages. High-level languages have been designed for ease of use and because they use ordinary words with their usual meanings. Machine code, on the other hand, is abstract, relatively difficult to write in the form of a program, but much more rapid and efficient in using storage space. Most commercial games programs are extremely complex and have to be written in machine code in order for the computer to operate sufficiently quickly and so that they may be contained in the available space. Some machines, such as the Hewlett-Packard HP-85, have only BASIC available: it is impossible to write programs for this computer in machine code even though the computer invisibly does so in translating and executing BASIC instructions. For most computers the translator is built-in and operates automatically so that the user may be unaware of its existence.

Certainly there is no need to know the detail of how it works. In fact, it can work in one of two ways. First, the machine may 'compile' the BASIC program. This is a once-for-all translation into machine code and the program is then stored as a machine code version for later use. This is an unusual approach in small computers. Secondly, the machine may 'interpret' the BASIC program. In this method each BASIC instruction is translated into machine code every time it is used. This is a very common method but it is obviously much slower in operation than a compiled program.

It is hard to imagine how the various functions that a computer performs can be achieved by simple operations such as adding and subtracting. However, this is generally what happens. For example, when the computer is instructed in BASIC to find the cosine of a stored number $(Y = COS(X))$ it does so by using a series summation. Other functions such as multiplication and division are also achieved by a number of very simple instructions.

Numbers and program instructions are both stored as binary numbers. The amount of storage space in a computer is important because it limits the scale of operations. If the Apple computer is taken as an example this will enable the character and size of storage capability to be envisaged. The 6502 processor used in the Apple is designed to process 8-bit binary numbers (called bytes). These numbers range in decimal notation as integers from 0 to 255. There is storage capacity for 64 k of these numbers (1 k or kilobyte, is 1024 or 2^{10} bytes). Each of these storage locations has an individual address for postage and retrieval. It is easy to work out that it requires two bytes to define the address (sequentially numbered) of a particular byte or memory. However, some of the 64 k of storage locations are used by the system management of the computer to hold programs which must be constantly available and which should not be lost when the machine is switched off.

The translation program, mentioned above, is one such program to be stored in this way; another is the set of standard mathematical programs for calculating cosines, logarithms, etc. There is, in addition, the all-important AUTOSTART program which gets the machine organized for work when it is switched on. If the analogy is made with the human worker waking up in the morning, this program gets him out of bed, washed, shaved, dressed, fed, on the train and then seated at his desk ready for the first instruction. As with many managers this section of storage is never wrong, it can never be criticized and it can never be sacked! It is permanently housed in a chip called a ROM (Read-Only Memory). In the Apple used by the author this occupies one-quarter of the memory space (16 k). The remaining 48 k of addressable storage is 'volatile'. This remainder is on a RAM chip (Random-Access Memory) and it is possible both to read and to change the contents. The contents of RAM storage space is lost when the machine is switched off, hence the use of the term volatile. The RAM space can be used by BASIC programs to store not only numbers but also instructions. In using BASIC there is no need to worry about which is which because the system management sorts it all out.

BASIC instructions are usually stored as text with each symbol occupying one byte according to the ASCII code. This code makes use of the integers between 0 and 255 to represent upper and lower case letters, numbers and a range of other symbols some of which are available on a keyboard, see Table 1.1. The byte itself can be represented in writing either as a decimal number or as two

Table 1.1 ASCII* codes for the common keyboard characters.

Decimal X	Hex code	Character CHR$ (X)	Decimal X	Hex code	Character CHR$ (X)	Decimal X	Hex code	Character CHR$ (X)
10	0A	LF	51	33	3	72	48	H
13	0D	CR	52	34	4	73	49	I
32	20	SP	53	35	5	74	4A	J
33	21	!	54	36	6	75	4B	K
34	22	"	55	37	7	76	4C	L
35	23	#	56	38	8	77	4D	M
36	24	$	57	39	9	78	4E	N
37	25	%	58	3A	:	79	4F	O
38	26	&	59	3B	;	80	50	P
39	27	'	60	3C	<	81	51	Q
40	28	(	61	3D	=	82	52	R
41	29	)	62	3E	>	83	53	S
42	2A	*	63	3F	?	84	54	T
43	2B	+	64	40	@	85	55	U
44	2C	,	65	41	A	86	56	V
45	2D	–	66	42	B	87	57	W
46	2E	.	67	43	C	88	58	X
47	2F	/	68	44	D	89	59	Y
48	30	0	69	45	E	90	5A	Z
49	31	1	70	46	F			
50	32	2	71	47	G			

* American Standard Code for Information Interchange. The Sinclair ZX-81 is more or less alone in using a different code. The table is, in fact, the so-called 7-bit subset because the first of the eight bits is zero. The full 8-bit set includes lower-case letters

hexadecimal digits (Hex for short). Hex code uses the symbols 0 to 9 followed by A to F to represent the sixteen possible values of a four-bit number from decimal 0 to 15 in value. The byte is split into two four-bit pieces and coded accordingly. Thus decimal 77 in binary digits is 01001101 and in Hex is 4D. This number also represents the letter M in ASCII code. A typical BASIC instruction, as text, requires about thirty bytes in memory. When translated into machine code it also occupies about thirty bytes. Machine code instructions occupy between one and three bytes. Numbers usually occupy five bytes and are stored in an exponential or floating point form. The number is stored in the form $m2^n$ where m occupies part of the 40 bits of space and n occupies the rest according to some predefined convention. When numbers are printed out as results from a BASIC program they often appear in the exponential form

$$2.4468E-8$$

This is a convenient way to present very large or very small numbers, the E indicating that the number which follows it is the power of ten to which the preceding number is to be raised. Numbers can also be input in this form.

Some machines (Apple for example) also have an INTEGER BASIC available as a separate high-level language with instructions similar to those of ordinary BASIC. It generally works more quickly because numbers occupy only two bytes but it is limited in usefulness. It will not be used for any programs in this book.

1.2 Peripherals

The actual cost of a 6502 microprocessor chip is about £5 and yet a complete microcomputer system will cost about two or three hundred times as much. Why should this be? The reason is partly that there must also be a provision of storage space and a decent box to put it all in. However, this accounts for only a small part of the total cost. The remainder is required for peripherals. These are other machines, mostly electro-mechanical, which enable the user to communicate with his computer. The computer communicates with the peripherals through 'interfaces' and these must have not only their own addresses but also the necessary matching circuitry to communicate properly in both inward and outward directions. This is usually achieved with an interface board which plugs into a 'port' on the main computer board (called the motherboard) and is then wired to the peripheral machine. These peripherals would typically comprise:

(1) an ASCII coded keyboard;
(2) a Visual Display Unit (VDU);
(3) a Cassette or Disc recorder;
(4) a Printer.

When the computer is switched on it will usually 'default' to the keyboard for input and the VDU for output. If contact with other peripherals is required it must be specifically requested by means of special instructions. The transmission of information to and from peripherals can be at different rates. These are usually set at the maximum rate possible by the interface board, often by means of manual switches. The transmission rate is measured in 'Baud', a unit of about one bit per second. For example, teleprinters usually operate at 120 Baud. This will transmit ASCII coded 8-bit characters at a rate of about ten per second. This is rather slower than might be expected because each 8-bit character has an additional two or three bits appended to check for start, stop and errors in transmission. Transmission can be in serial form (one bit at a time) or parallel form (one byte or eight bits at a time). There are international standards for these two types of transmission called RS 232 for serial transmission and IEE 488 for parallel transmission.

Computers are often connected directly to industrial processes or scientific experiments for purposes of measurement, analysis and control. If the signals to and from the computer are variable electrical voltages then special interfaces are required called analogue-to-digital converters (ADCs), or their converse, digital-to-analogue converters (DACs). These will convert ordinary electrical voltages into binary coded digital numbers and vice versa. Straightforward BASIC programs can then be written so that the computer can measure and control these processes.

1.3   Execution of BASIC programs

The first step in running a BASIC program is to ensure that the computer contains the sequence of instructions in the required numerical order of execution. This can be achieved either by typing in the instructions one by one, with line numbers in the required order, or by using the LOAD instruction (OLD on some computers) to retrieve an existing program from tape or disc. When the program is installed it can be executed by the RUN instruction. The program is executed in sequence until there is a natural completion (STOP or END instruction) or an error causes a halt. In the latter case an error message is usually displayed. In all of the foregoing, the master programs within the machine have interpreted certain words in specific ways. This happens throughout the control and execution of a program and if spelling errors are made the computer indicates this by the output of an error message. There can be gaps in the line numbers which the computer will ignore though it will always perform instructions in numerical order. Indeed, it is useful to have gaps in the line number sequence so that additional instructions and corrections can be made. In this book all programs have line intervals of ten for these purposes. When a program is deemed to be satisfactory it can be permanently stored using the SAVE instruction which will place it on tape or disc (where such peripherals are installed). If this is not done the program will be lost when the computer is switched off. Instructions such as LOAD, SAVE and RUN do not have line numbers because they are obeyed immediately (called immediate execution). The instructions in the program are only obeyed in their due turn and are therefore called deferred execution instructions. Some instructions, such as PRINT, can be used in either way but they will always require a line number if execution is to be deferred.

1.4   BASIC programs

Any program is an ordered set of instructions which is designed to achieve a desired result by a pre-defined sequential process. These instructions have to be written in a language or code and in this book BASIC is used. This section gives a brief summary of BASIC words with their meaning and use. The examples in the main body of the book illustrate extensively how the language is used. For a more detailed description of BASIC see Further reading, p. 115, ref. 2.

1.4.1   Arithmetic instructions

The simplest kind of instruction, one which performs arithmetic, is of the form

10 LET A = B + C

Line numbering has already been discussed. The instruction itself has the following meaning: Retrieve the numbers stored in the locations labelled B and C, add them together and store the result in the location labelled A. The equality symbol has the meaning 'become equal to'. B or C could equally well be numbers or mathematical expressions. The word LET is often omitted in typing the instruction into the computer although it may appear automatically when the program is listed on the VDU or printer for checking. The function to the right of the equality symbol can contain a wide range of operations other than $+$, $-$, $*$ (multiply) and $/$ (divide). A list (not necessarily complete) with X representing a number, variable or expression is as follows:

SQR(X)   Square root of X
ABS(X)   Absolute value of X
SGN(X)   Sign of X; 1 if $X > 0$, 0 if $X = 0$, -1 if $X < 0$
INT(X)   Largest integer not greater than X
EXP(X)   e (the root of natural logarithms) to the power X
LOG(X)   Natural logarithm of X
SIN(X)   Sine of (X in radians)
COS(X)   Cosine of (X in radians)
TAN(X)   Tangent of (X in radians)
ATN (X) Inverse tangent of X (in radians between $-\pi/2$ and $\pi/2$)
RND(X)   A random number between 0 and 1

The raising of a number to a power requires the symbol ↑, thus

10 LET A = B ↑ 3.5

This will place B raised to the power 3.5 in location A.

There is an order of priority for the execution of compounded arithmetical operations so as to avoid ambiguity, as follows:

( )   brackets
↑   exponentiation
* /   multiplication or division
+ −   addition or subtraction

In writing instructions any doubt about possible ambiguity can be removed by using brackets.

1.4.2   Variables

In the previous section A, B, C and X have been used to represent variable quantities, as in ordinary algebra. They are, in fact, nothing more than a label for the address of the memory space which the number occupies. A second kind of variable is the array or subscripted variable. This is particularly useful for vibration calculations because the array may be used to represent a matrix or vector, see Chapter 5. For matrix work arrays are one- or two-dimensional and each element will be numbered. For example, A(6) represents the sixth item in an array, list or vector called A. Arrays are described by a single letter and must have memory space allotted before use in a program by means of a DIM(ension) statement, for example

    10 DIM A(10)
    20 DIM B(4,6)

B has been dimensioned to make space for a 4 × 6 matrix.

A third kind of variable is the 'string' variable consisting of text and other symbols. The name of the variable consists of a letter followed by the \$ symbol. String variables can also be arrays and they then have to be dimensioned as for numeric variables, for example

    30 DIM C\$(20)

This instruction makes space available for twenty different strings.

1.4.3   Control instructions

It is often necessary to make branches in the normal sequence of following instructions. This can be done unconditionally by using a GOTO instruction, for example

    10 GOTO 100

This will cause the program to skip over all line numbers between 11 and 99 inclusive. The same thing can also be done, subject to a condition, by the instruction

10 IF (conditional statement) THEN 100

The conditional statement can take a variety of forms involving numeric and string variables and either equalities or inequalities. If the statement is true then the branch will occur, otherwise control will proceed to the next instruction.

Another widely used feature of programs is the loop. This is used when a set of instructions is to be repeated. For this purpose an IF ... THEN instruction, in conjunction with a loop counter, can be used but it is more usual to write a FOR ... NEXT sequence. This is illustrated in the following example of a segment of program which will calculate COS(X) to the power 7 and place it in location B

```
10 LET B = 1
20 FOR I = 1 TO 7
30 LET B = B * COS(X)
40 NEXT I
```

Note that every FOR must have a corresponding NEXT. Such loops can be nested one inside another.

1.4.4 Subroutines

A subroutine is a set of instructions which may be required repetitively throughout a program. Rather than write the set out each time it is required, it is easier and more efficient to have a subroutine which is called up by means of a GOSUB control instruction. For example, in matrix vibration calculations there is a constant need to be able to multiply matrices (arrays) and this can easily be done with a subroutine (see Chapter 5). The GOSUB instruction must specify the starting line number of the subroutine. All the variables which the subroutine expects to find must be correctly set before entry. The final instruction of the subroutine itself must be RETURN. This will effectively GOTO the instruction line following the GOSUB instruction.

1.4.5 Input/output

These instructions constitute the means of acquiring data at the start of the program and of displaying the final results. In many cases the instructions involve use of the peripherals, such as a printer. The input instructions are INPUT and READ. When INPUT is encountered the computer will wait until appropriate data are available from the connected input peripheral, usually the keyboard. More than one quantity can be INPUT by this instruction and it will then have the form

    10 INPUT A,B,C

At execution the computer will output a question mark to prompt the user to reply with three numbers separated by commas and terminated by carriage return. The READ instruction, in similar form to the INPUT instruction above, will take numbers from the first DATA statement in the program, for example

    600 DATA 10,12,14

Any subsequent READ instruction will take data from the same or following DATA statements sequentially. If it is required to reset the taking of data to the first data line then the RESTORE instruction must be used.

Output of results and of text is achieved with the PRINT instruction. Typical PRINT instructions are as follows:

    10 PRINT
    20 PRINT X
    30 PRINT X$
    40 PRINT X; "METRES"

Instruction 10 will simply output a linefeed to the printer or VDU. Instruction 20 will output the current value of variable X, instruction 30 the current string in X$. Instruction 40 shows two additional features: first the use of the semicolon as a delimiter between two items of printout causes them to be printed without a gap on the same line. A comma would cause tabulation every 15 columns. A semicolon at the end of the instruction eliminates the output of carriage return/linefeed. Secondly, the word METRES is faithfully mimicked at the printer because it is enclosed in quotes.

1.4.6 Special functions

REM (Remark): An instruction starting with this word is disregarded in execution but allows the programmer to insert useful comment into the program.

DEF FN (Define Function): This instruction is rather like a one-line subroutine. A simple example will illustrate its use. Suppose that it is wished to calculate the volume of a sphere $(4\pi r^3/3)$ at several points in a program. The DEF FN is used as follows:

    10 DEF FNV(R) = (4 * PI * R ↑ 3) / 3

Subsequently the variable FNV(X) will take the value of the volume of the sphere of radius X. Note that the use of the variable PI for π is not available on all computers.

CHR$(X): This is used to output the ASCII symbol corresponding to the decimal value X, see Table 1.1. The value X must lie between 0 and 255. In fact, it is a grossly inefficient way to output symbols which appear on the keyboard and is therefore usually reserved for all sorts of clever purposes such as controlling the typeface on the printer, ringing the warning bell, etc.

PEEK and POKE: These instructions are used to inspect and insert contents into individual single byte memory locations. The POKE instruction, like burying a dead elephant, should not be undertaken lightly, because the program may easily be wrecked. PEEK and POKE are not available on BBC computers.

1.4.7   Checking and editing

All programs need to be checked before they are RUN to see that no typing errors or omissions have been made. Errors in syntax, if undetected at this stage, will lead to an error message at RUN time. Other errors or omissions may not be detected and will simply lead to incorrect results. It is important, therefore, that one or two check calculations are made by hand so that the program can be verified. In complicated programs this may not always be easy. In such cases it may be possible to check the program in sections.

If errors are found they have to be corrected by editing the program. This is easily done by retyping the offending line using the same line number. Make sure that the edited program is immediately SAVEd so that the copy stored on cassette or disc is also brought up-to-date.

1.4.8   Variations between computers

Users of this book will have a variety of computers at their disposal ranging from large mainframe machines through minicomputers, such as PDP-11 and DGC Nova, down to microcomputers, such as Apple, PET, BBC and Sinclair. There are even battery portables available such as the Epson HX-20. The information given in this chapter is a minimum BASIC kit which will be useful on all these machines. The programs in the following chapters make use of this minimum set of instructions (unless specifically stated otherwise) so that all readers, whether the engineer working for a living or a student at home in front of his TV set, should find the programs useful as they stand. It may be noted that some of the larger machines may have extended BASIC languages installed and these can be particularly useful for vibration analysis because they often include special matrix instructions.

It is unfortunate that there is no uniformity in the programming of graphics. For this reason none of the programs in this book produces graphical output. In computer-based engineering design analysis it is very useful to have graphical output of results so that there can be a visual impact when small design changes are made. Design is a subject where the interaction between man and computer leads very rapidly, by an iterative process, to results which may otherwise take a long time to achieve. The reader is therefore encouraged to explore the possibility of producing graphical output of results from programs in this book. He will find that this can be a rewarding and educational achievement.

Chapter 2

Simple systems with no damping

This chapter describes the vibration analysis of engineering systems which may be modelled by a single degree of freedom, as shown in Figure 2.1. This can be an adequate model for many real systems even though they may have many degrees of freedom.

There are also circumstances in which the damping of vibration is small and unimportant, for example, (1) in the determination of natural frequency in relation to the running speed of a machine, or (2) in the determination of a peak stress in a system following the application of an impulsive blow. Thus, although damping always exists in practice, it is not considered in this chapter. The wide range of problems for which damping is important will be covered in the next chapter.

ESSENTIAL THEORY

2.1 Equation of free motion

Referring to Figure 2.1, with the mass m displaced by x, the application of Newton's second law of motion gives the equation of motion

$$m\ddot{x} + kx = 0 \tag{2.1}$$

This has a general solution

$$x = A \cos \omega_n t + B \sin \omega_n t \tag{2.2}$$

or

$$x = M \cos(\omega_n t + \phi) \tag{2.3}$$

in which A and B (or M and ϕ) are constants determined by initial conditions. More importantly the natural circular frequency, ω_n, is given by

$$\omega_n = \sqrt{k/m} \tag{2.4}$$

The cyclic frequency f_n (Hz) is given by $\omega_n/2\pi$. The amplitude of the

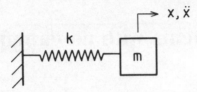

Figure 2.1 Model of the simple system with one degree of freedom

oscillation is given by

$$|x| = M = (A^2 + B^2)^{1/2}$$

It is often the most important quantity in design calculations. Values for both velocity and acceleration responses may be derived by differentiation

$$\dot{x} = -\omega_n A \sin \omega_n t + \omega_n B \cos \omega_n t$$

$$\ddot{x} = -\omega_n^2 A \cos \omega_n t - \omega_n^2 B \sin \omega_n t$$

with corresponding magnitudes

$$|\dot{x}| = \omega_n (A^2 + B^2)^{1/2} = \omega_n M$$

$$|\ddot{x}| = \omega_n^2 (A^2 + B^2)^{1/2} = \omega_n^2 M$$

2.2   Conservation of energy

The acceleration $\ddot{x}$ may be expressed as $v\,dv/dx$ where the speed v is the same as $\dot{x}$. Thus Equation (2.1) may be written

$$mv\,dv + kx\,dx = 0$$

and integration gives

$$mv^2/2 + kx^2/2 = E \quad \text{(a constant)} \tag{2.5}$$

This equation is an expression of the principle of conservation of energy. The first term is the instantaneous value of the kinetic energy of the mass m whilst the second term is the instantaneous value of the potential energy or strain energy stored in the spring. Thus the total energy of the system, E, remains constant and during motion there is an oscillatory interchange of kinetic and potential energies. When the mass m is at a point of furthest excursion E is entirely in the form of strain energy, whilst when the mass is passing through the neutral or central position, E is entirely in the form of kinetic energy. This physical view of vibrations is a powerful one and is the basis of an approximate method of determining natural frequencies called

Rayleigh's method (see Chapter 6). It also provides a powerful method of analysing vibrations of mechanisms and linkages.

2.3 Forced motion: resonance

When a variable external force $F(t)$ is applied to the mass in Figure 2.1 the equation of motion becomes

$$m\ddot{x} + kx = F(t) \tag{2.6}$$

It is both convenient and useful to analyse this equation for $F(t) = F_0 \cos \omega t$, not only because the solution is easy to find but also because many real force histories can be analysed as a series of sinusoidal functions by making use of Fourier Series (see Chapter 3). In many real cases the function $F(t)$ is sinusoidal anyway. The solution to the equation is

$$x = A \cos \omega_n t + B \sin \omega_n t + \frac{F_0}{(k - m\omega^2)} \cos \omega t \tag{2.7}$$

Once again A and B are to be determined from the initial conditions. The final term is called the steady state solution, for reasons which are explained in the next chapter. It represents motion at the forcing frequency as distinct from motion at the natural frequency. The amplitude of the steady state motion is

$$\frac{F_0}{k} \frac{1}{\left[1 - \left(\dfrac{\omega}{\omega_n}\right)^2\right]} \tag{2.8}$$

F_0/k represents the value of x for a static force of F_0. The term by which it is multiplied is called the Dynamic Magnification Factor or DMF for short. It can be seen that as ω approaches ω_n the value of the DMF becomes very large and this phenomenon is known as resonance. A graph of DMF against frequency (see Figure 2.2) is known as the frequency response function of the system.

2.4 Earth motion

In some practical instances motion in a vibrating system is induced not by an applied force but by motion of the ground on which the system rests. Such is the excitation of vehicles travelling over rough ground or of buildings in an earthquake. Such also is the excitation of the standard type of seismic vibration measuring instrument which consists of a system exactly like that shown in Figure 2.1 with the addition of a transducer fixed between ground and mass to measure their relative motion. Instruments of this kind can be designed so that

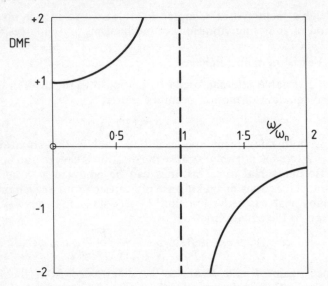

Figure 2.2 Frequency response function for the simple system

the transducer, within a limited frequency range, will effectively measure ground displacement (Seismometer), ground velocity (Vibrometer) or ground acceleration (Accelerometer).

If the symbol x is retained for the absolute motion of the mass and the symbol x_0 is used for the absolute ground motion, the equation of free motion is

$$m\ddot{x} + k(x - x_0) = 0$$

It is convenient to form a new variable $y = (x - x_0)$ which is the relative deflection between the mass and earth. This is an appropriate variable for (1) buildings subject to earthquakes, (2) vehicles traversing rough ground, and (3) seismic instruments. For in case (1) it relates directly to stress in the building, and in case (2) is relates directly to the dynamic load in the suspension system. For seismic instruments y is the variable which the transducer will measure. The differential equation in terms of y is

$$m\ddot{y} + ky = -m\ddot{x}_0 \tag{2.9}$$

This equation is now similar to Equation (2.6) but with the force $F(t)$ replaced by the force which would be experienced by the mass m if it were rigidly attached to ground. The solution to the equation is similar to that obtained in the last section except that here it will be for relative, rather than absolute, motion.

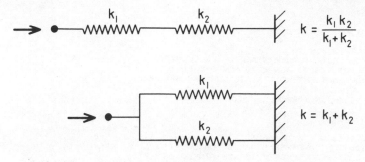

Figure 2.3 Properties of springs in series and parallel

WORKED EXAMPLES

Example 2.1   Compound spring stiffnesses

Write a program to calculate the nett effective stiffness of a group of springs in series and parallel. Make use of the results shown in Figure 2.3.

```
10   REM  EX2POINT1
20   PRINT "COMPOUND SPRING STIFFNESSES"
30   PRINT "-----------------------------"
40   PRINT
50   PRINT "DO YOU WANT A CALCULATION FOR TWO SPRINGS"
60   PRINT "         IN SERIES ?     (1)"
70   PRINT "         IN PARALLEL ?   (2)"
80   INPUT N
90   PRINT "INPUT K1"
100  INPUT K1
110  PRINT "INPUT K2"
120  INPUT K2
130  IF N = 1 THEN 160
140  K3 = K1 + K2
150  GOTO 170
160  K3 = K1 * K2 / (K1 + K2)
170  PRINT "RESULT IS K = ";K3
180  PRINT "DO YOU WANT ANOTHER STAGE OF CALCULATION?"
190  GET C$
200  IF C$ = "N" THEN 280
210  K1 = K3
220  PRINT "THE LAST RESULT FOR K IS NOW STORED AS K1"
230  PRINT "DO YOU WANT THE NEXT STAGE FOR TWO SPRINGS"
240  PRINT "         IN SERIES ?     (1)"
250  PRINT "         IN PARALLEL ?   (2)"
260  INPUT N
270  GOTO 110
280  STOP

JRUN
COMPOUND SPRING STIFFNESSES
--------------------------

DO YOU WANT A CALCULATION FOR TWO SPRINGS
         IN SERIES ?     (1)
         IN PARALLEL ?   (2)
?2
```

```
INPUT K1
?10
INPUT K2
?10
RESULT IS K = 20
DO YOU WANT ANOTHER STAGE OF CALCULATION?
THE LAST RESULT FOR K IS NOW STORED AS K1
DO YOU WANT THE NEXT STAGE FOR TWO SPRINGS
          IN SERIES ?    (1)
          IN PARALLEL ?  (2)
?1
INPUT K2
?10
RESULT IS K = 6.66666667
DO YOU WANT ANOTHER STAGE OF CALCULATION?
THE LAST RESULT FOR K IS NOW STORED AS K1
DO YOU WANT THE NEXT STAGE FOR TWO SPRINGS
          IN SERIES ?    (1)
          IN PARALLEL ?  (2)
?2
INPUT K2
?10
RESULT IS K = 16.6666667
DO YOU WANT ANOTHER STAGE OF CALCULATION?

BREAK IN 280
```

Program notes

(1) The problem solved here is to find the nett stiffness of the spring set in Figure 2.4(a). Each spring has a stiffness of 10 units and the arrow indicates the point for which nett stiffness is required.

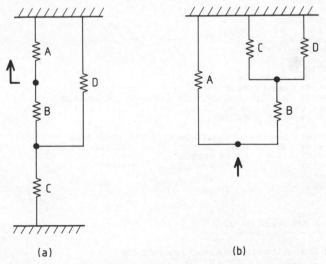

(a)   (b)

Figure 2.4 Sample problem for Example 2.1 with preparation for computing

The first two stages, before using the program, are

(a) To reduce this diagram to an equivalent series/parallel diagram. The determination of this diagram is often the most difficult part of the process. For this case it is shown in Figure 2.4(b).
(b) To decide on the order of the computer solution. In this case it is as follows
    (i)   C + D in parallel: Result K = 20
    (ii) (C + D) + B in series: Result K = 6.67
    (iii) ((C + D) + B) + A in parallel: Result K = 16.67
    (iv) BREAK

(2) The program is under user control by means of the GET instruction in line 190. This takes one character from the keyboard into the string C\$. The expected reply is either "Y" (YES) or "N" (NO). The GET instruction can also be used to provide a pause. Throughout this book multiple options are numbered and an appropriate reply is required from the keyboard, see lines 50 to 80.

Example 2.2   Natural frequency calculation

Natural frequencies may be easily calculated using Equation (2.4), given the values for spring stiffness and mass in consistent units. It is also a remarkable fact that the natural frequency of a simple system is directly related to the static deflection, d, of the mass under its own weight. For it is clear that $d = mg/k$, where g is the acceleration due to gravity, so that

$$\omega_n = \sqrt{g/d}$$

A branch of the program therefore calculates the result for cases where only this information is known accurately (for example, by experiment). It is also useful to be able to give the result in either radians/second (rad/s), cycles per second (Hz) or revolutions per minute (rev/min). In practice the spring may be one of a variety of kinds: a coil spring, a beam in bending or a flexible block or link. The program is checked for the machine which is the subject of Example 2.3 for which the mass is 220 kg and the spring stiffness is 35 kN/m.

```
10   REM   EX2POINT2
20   PRINT "NATURAL FREQUENCY CALCULATION"
30   PRINT "SIMPLE SYSTEM, NO DAMPING"
40   PRINT "------------------------------"
50   PRINT
60   PRINT
70   PRINT "    USE CONSISTENT UNITS"
80   PRINT
90   PRINT
100  PRINT "IS YOUR CALCULATION TO BE BASED ON"
```

```
110   PRINT "  MASS AND STIFFNESS VALUES ?   (1)"
120   PRINT "  STATIC DEFLECTION VALUES ?    (2)"
130   INPUT I
140   IF I = 2 THEN 210
150   PRINT "INPUT THE MASS VALUE"
160   INPUT M
170   PRINT "INPUT THE STIFFNESS VALUE"
180   INPUT K
190 W = SQR (K / M)
200   GOTO 260
210   PRINT "INPUT THE STATIC DEFLECTION VALUE"
220   INPUT D
230   PRINT "INPUT THE VALUE OF GRAVITY"
240   INPUT G
250 W = SQR (G / D)
260   PRINT "DO YOU WANT THE RESULT IN"
270   PRINT "   RADIANS PER SECOND ?          (1)"
280   PRINT "   REVOLUTIONS PER MINUTE ?     (2)"
290   PRINT "   CYCLES PER SECOND (HZ) ?     (3)"
300   INPUT I
310   PRINT
320   PRINT "NATURAL FREQUENCY = ";
330   IF I = 2 THEN 370
340   IF I = 3 THEN 390
350   PRINT W;" RADIANS PER SECOND"
360   GOTO 400
370   PRINT (30 * W / 3.142);" REVOLUTIONS PER MINUTE"
380   GOTO 400
390   PRINT (W / 6.283);" HZ"
400   END


JRUN
NATURAL FREQUENCY CALCULATION
SIMPLE SYSTEM, NO DAMPING
------------------------------


    USE CONSISTENT UNITS

IS YOUR CALCULATION TO BE BASED ON
   MASS AND STIFFNESS VALUES ?   (1)
   STATIC DEFLECTION VALUES ?    (2)
?1
INPUT THE MASS VALUE
?220
INPUT THE STIFFNESS VALUE
?35000
DO YOU WANT THE RESULT IN
   RADIANS PER SECOND ?          (1)
   REVOLUTIONS PER MINUTE ?      (2)
   CYCLES PER SECOND (HZ) ?      (3)
?3

NATURAL FREQUENCY = 2.00750032 HZ
```

Program notes

(1) As written, the program can be used for any unit system. If it is restricted to SI units lines 230 and 240 will become redundant if a value of 9.81 is used for G in line 250.

(2) This is the first example in this book which highlights the fact that a variety of units are often used for the same quantity. Here it is

natural frequency which may be expressed in one of three widely used forms. It would also be easy to offer a further alternative; the natural period of oscillation, given algebraically by $2\pi/W$.

Example 2.3   Forced harmonic motion

Write a program to determine the amplitude of forced harmonic motion given the force amplitude, spring stiffness, system mass and frequency of excitation. Express the result as either displacement, velocity or acceleration magnitude. Use the program to solve the following problem.

An engine of mass 220 kg rests on a spring of stiffness 35 kN/m. There is a vertical oscillating force, of amplitude 100 N, acting on the engine. It is known that operation of the carburettor becomes faulty if vertical accelerations exceed 0.9 m/s^2. Explore the speed range 500 to 3000 rev/min to see if the selected mounting spring is satisfactory.

```
10   REM EX2POINT3
20   PRINT "FORCED HARMONIC MOTION"
30   PRINT "SIMPLE SYSTEM, NO DAMPING"
40   PRINT "------------------------"
50   PRINT
60   PRINT
70   PRINT "    USE CONSISTENT UNITS"
80   PRINT
90   PRINT
100  PRINT "INPUT BASIC DATA FOR THE PROBLEM AS FOLLOWS"
110  PRINT
120  PRINT "FORCE AMPLITUDE ?"
130  INPUT F
140  PRINT "SPRING STIFFNESS ?"
150  INPUT K
160  PRINT "SYSTEM MASS ?"
170  INPUT M
180  DS = F / K
190  WN =  SQR (K / M)
200  PRINT "DO YOU WANT SOLUTIONS FOR"
210  PRINT "   A RANGE OF FREQUENCIES ? (1)"
220  PRINT "   A SINGLE FREQUENCY ?       (2)"
230  INPUT I
240  IF I = 2 THEN 290
250  PRINT "INPUT START POINT, END POINT, INCREMENT"
260  PRINT "FOR THE FREQUENCY RANGE, IN HZ"
270  INPUT R,S,T
280  GOTO 330
290  PRINT "INPUT THE FREQUENCY VALUE IN HZ"
300  INPUT R
310  S = R
320  T = 1
330  PRINT "DO YOU WANT RESPONSE VALUES FOR"
340  PRINT "    DISPLACEMENT ?     (1)"
350  PRINT "    VELOCITY ?         (2)"
360  PRINT "    ACCELERATION ?     (3)"
370  INPUT I
380  PRINT
390  PRINT "FREQUENCY","RESPONSE"
400  PRINT "---------","--------"
410  FOR F = R TO S STEP T
```

```
420  PRINT F,
430  W = 6.283 * F
440  D = 1 - (W * W / (WN * WN))
450  IF D = O THEN 540
460  H = 1 / D
470  IF I = 1 THEN 520
480  IF I = 2 THEN 510
490  H = H * W * W
500  GOTO 520
510  H = H * W
520  PRINT (H * DS)
530  GOTO 550
540  PRINT "INFINITY"
550  NEXT F
560  PRINT "----------","---------"
570  END
JRUN
FORCED HARMONIC MOTION
SIMPLE SYSTEM, NO DAMPING
-------------------------

    USE CONSISTENT UNITS

INPUT BASIC DATA FOR THE PROBLEM AS FOLLOWS

FORCE AMPLITUDE ?
?100
SPRING STIFFNESS ?
?35000
SYSTEM MASS ?
?220
DO YOU WANT SOLUTIONS FOR
    A RANGE OF FREQUENCIES ?  (1)
    A SINGLE FREQUENCY ?      (2)
?1
INPUT START POINT, END POINT, INCREMENT
FOR THE FREQUENCY RANGE, IN HZ
?10,50,10
DO YOU WANT RESPONSE VALUES FOR
    DISPLACEMENT ?    (1)
    VELOCITY ?        (2)
    ACCELERATION ?    (3)
?3
```

FREQUENCY	RESPONSE
10	-.473633143
20	-.459171675
30	-.456589993
40	-.455693249
50	-.455279376

Program notes

(1) This program makes use of Equation (2.8). Line 180 calculates the Static Deflection (DS) under the action of the force. Line 460 calculates the Dynamic Magnification Factor.

(2) The possibility of resonant excitation has to be allowed for and for a system with no damping an infinite value for response will occur. This difficulty is resolved by the branch at line 450.

(3) In the solution of a variety of problems of this kind response

values may be required either in terms of displacement, velocity or acceleration. These options are allowed in lines 330 to 370 inclusive. In the problem solved here, acceleration is the important response quantity. The program may be elaborated by indicating, in the printout, whether the response values are accelerations, velocities or displacements.

(4) The program allows for either a point frequency solution or for a range of frequencies (lines 200 to 230 inclusive). The latter permits a frequency response curve to be plotted and, on suitable computers, this could lead to a graphical output.

(5) The basic data here have been expressed in SI units and thus the response values are of acceleration amplitudes in m/s^2. As all of these are less than $0.9\,m/s^2$ the design is satisfactory. The response values are all negative, indicating that the machine is being excited above its natural frequency, which, from Example 2.2, is known to be about 120 rev/min (2 Hz).

(6) Note that W * W is preferred to W ↑ 2 in line 440, and elsewhere, as it is quicker and more accurate.

Example 2.4   Data analysis from a bridge experiment

A motorway overbridge is vibrating freely in its fundamental mode of vibration as a result of excitation by passing traffic. Values of displacement and velocity at the centre are simultaneously measured using suitable instruments connected to a computer interface. The values are digitally sampled at short, regular time intervals of unknown length. They are stored first at locations D1 and V1 respectively. These values are then transferred to locations D2 and V2 respectively before the next set is stored again at D1 and V1. During the sampling interval the computer has time to run 40 lines of BASIC program. Write a program of 40 lines or less to process these data and printout the amplitude of motion, the vibration frequency and the energy of the system. The stiffness of the bridge at the centre is known from a static load test to be 1000 N/mm. To test the program four consecutive values of displacement and velocity are as follows:

D (mm)	V (mm/s)
5.4	−251
−17.1	−134
−14.5	179
9.4	230

A little theoretical development is required to define how data are to be processed to obtain the required result. Starting from Equation (2.3) the instantaneous value for velocity is obtained:

$$\text{variable } D = x = M \cos(\omega_n t + \phi)$$

$$\text{variable } V = dx/dt = -\omega_n M \sin(\omega_n t + \phi)$$

Thus at any instant if M is constant

$$D^2 + \frac{1}{\omega_n^2} V^2 = M^2$$

If D and V are known at two consecutive instants then two equations are available for the unknowns $1/\omega_n^2$ and M^2. These can be solved to give:

$$\omega_n^2 = \frac{(V2)^2 - (V1)^2}{(D1)^2 - (D2)^2}$$

and M^2 follows directly. Thus, the displacement amplitude M and the natural frequency ω_n can be calculated. The total energy of the system, E, is calculated using Equation (2.5)

$$E = kD^2/2 + mV^2/2 = k\left(D^2 + \frac{1}{\omega_n^2} V^2\right)\bigg/2 = kM^2/2$$

```
10   REM EX2POINT4
20   PRINT "VIBRATION DATA ANALYSIS FOR A BRIDGE"
30   PRINT "---------------------------------------"
40   PRINT "THE CORE PROGRAM WHICH IS USED FOR DATA"
50   PRINT "ANALYSIS STARTS ON LINE 160. THE LINES"
60   PRINT "OF PROGRAM BEFORE THIS SET UP DATA AS"
70   PRINT "IT WOULD BE IN AN EXPERIMENT"
80   PRINT
90   PRINT "    USE CONSISTENT UNITS"
100   PRINT
110   PRINT "INPUT THE BRIDGE STIFFNESS K"
120   INPUT K
130   PRINT "INPUT D2, V2"
140   INPUT D2,V2
150   REM   START OF DATA PROCESSING PROGRAM SEGMENT
160   PRINT "INPUT D1, V1"
170   INPUT D1,V1
180   W = D2 * D2
190   X = V2 * V2
200   Y = D1 * D1
210   Z = V1 * V1
220   WS = (X - Z) / (Y - W)
230   MS = Y + Z / WS
240   E = K * MS / 2
250   PRINT "BRIDGE DISPLACEMENT AMPLITUDE = "; SQR (MS)
260   PRINT "BRIDGE NATURAL FREQUENCY = "; SQR (WS) / 6.283;" HZ"
270   PRINT "VIBRATION ENERGY = ";E
280   D2 = D1
290   V2 = V1
300   PRINT
```

```
310  PRINT "IS THERE ANOTHER DATA SET ?"
320  GET C$
330  PRINT
340  IF C$ = "Y" THEN 160
350  STOP
]RUN
```
VIBRATION DATA ANALYSIS FOR A BRIDGE

THE CORE PROGRAM WHICH IS USED FOR DATA
ANALYSIS STARTS ON LINE 160. THE LINES
OF PROGRAM BEFORE THIS SET UP DATA AS
IT WOULD BE IN AN EXPERIMENT

    USE CONSISTENT UNITS

INPUT THE BRIDGE STIFFNESS K
?1E6
INPUT D2, V2
?0.0054,-0.251
INPUT D1, V1
?-0.01/1,-0.134
BRIDGE DISPLACEMENT AMPLITUDE = .0199335813
BRIDGE NATURAL FREQUENCY = 2.08195839 HZ
VIBRATION ENERGY = 198.673831

IS THERE ANOTHER DATA SET ?

INPUT D1, V1
?-0.0145,0.179
BRIDGE DISPLACEMENT AMPLITUDE = .0199286263
BRIDGE NATURAL FREQUENCY = 2.08392052 HZ
VIBRATION ENERGY = 198.575073

IS THERE ANOTHER DATA SET ?

INPUT D1, V1
?0.0094,0.230
BRIDGE DISPLACEMENT AMPLITUDE = .0199369569
BRIDGE NATURAL FREQUENCY = 2.08207152 HZ
VIBRATION ENERG` = 198.74112⁵

3 THERE ANOTHER DATA SET ?

BREAK IN 350

Program notes

(1) The core program consists of 17 lines from 160 to 330 inclusive. The requirement for a core program of 40 lines or less is therefore satisfied. However, this is a rather artificial requirement because (a) some of the given lines are not strictly necessary, e.g. lines 300 and 320, and (b) some lines take longer to perform than others. In practice, either a timing limit is given or the sampling rate is adjusted to suit the time taken. Where it is necessary to have very rapid sampling the processing program must be written in machine code. On some computers, it is possible to speed up the processing by compiling the BASIC instructions rather than interpreting them (see Chapter 1).

(2) Note that all the basic data have been input in SI units and this

has led to the use of the so-called scientific notation for the input of the bridge stiffness. Thus, 1E6 has the same meaning as 1000000. The values obtained as results are moderately constant bearing in mind the rounding errors in the input data for D and V. Displacement amplitude are in metres, energy values in joules.

Example 2.5   Vibration isolation

Many machines, for example the engines of motor cars, are mounted on soft springs to minimize the transmission to ground of any alternating force arising within the machine. Referring to Figure 2.1 the force transmitted to ground is kx for any instantaneous value of x. The transmissibility, T, is the ratio of the amplitude of the force transmitted to ground to the amplitude of the applied force F_0. Thus, using Equation (2.8)

$$T = \frac{|kx|}{F_0} = \left| \frac{1}{(1 - (\omega/\omega_n)^2)} \right| \tag{2.10}$$

Write a program to calculate T for a range of running speeds for a machine of given mass and isolator of given stiffness. Check the running of the program for the machine described in Example 2.3 with $m = 220\,\text{kg}$ and $k = 35\,\text{kN/m}$ and for the given speed range of 600 to 3000 rev/min.

```
10   REM EX2POINT5
20   PRINT "VIBRATION ISOLATION"
30   PRINT "--------------------"
40   PRINT
50   PRINT "USE CONSISTENT UNITS"
60   PRINT
70   PRINT "INPUT MACHINE MASS"
80   INPUT M
90   PRINT "INPUT ISOLATOR STIFFNESS"
100   INPUT K
110  WN = SQR (K / M)
120   PRINT "IS THE RANGE OF RUNNING SPEEDS FOR"
130   PRINT "YOUR CALCULATION MEASURED IN"
140   PRINT "     RADIANS PER SECOND ?        (1)"
150   PRINT "     REVOLUTIONS PER MINUTE ?    (2)"
160   PRINT "     CYCLES PER SECOND ?         (3)"
170   INPUT I
180   IF I = 1 THEN 230
190   IF I = 2 THEN 220
200  WN = WN / 6.283
210   GOTO 230
220  WN = WN * 30 / 3.142
230   PRINT "INPUT START POINT, END POINT, INCREMENT"
240   PRINT "FOR THE RUNNING SPEED RANGE"
250   INPUT X,Y,Z
260   PRINT
270   PRINT "-------------","------------------"
280   PRINT "RUNNING SPEED","TRANSMISSIBILITY"
290   PRINT "-------------","------------------"
```

```
300  FOR W = X TO Y STEP Z
310  T =  ABS (1 / (1 - (W / WN) ^ 2))
320  PRINT W,T
330  NEXT W
340  PRINT "-------------","-----------------"
350  STOP
]RUN
VIBRATION ISOLATION
--------------------

USE CONSISTENT UNITS

INPUT MACHINE MASS
?220
INPUT ISOLATOR STIFFNESS
?35000
IS THE RANGE OF RUNNING SPEEDS FOR
YOUR CALCULATION MEASURED IN
        RADIANS PER SECOND ?      (1)
        REVOLUTIONS PER MINUTE ?  (2)
        CYCLES PER SECOND ?       (3)
?2
INPUT START POINT, END POINT, INCREMENT
FOR THE RUNNING SPEED RANGE
?600,3000,200

--------------   -----------------
RUNNING SPEED    TRANSMISSIBILITY
--------------   -----------------
600              .0419789887
800              .0231873272
1000             .0147170398
1200             .0101744135
1400             7.45495589E-03
1600             5.69774518E-03
1800             4.49654505E-03
2000             3.63909245E-03
2200             3.00561614E-03
2400             2.52434061E-03
2600             2.15011454E-03
2800             1.85337634E-03
3000             1.61411115E-03
--------------   -----------------

BREAK IN 350
```

Program notes

(1) The transmissibility (T) in line 310 is, in this case, the same as the DMF mentioned immediately after Equation (2.8) with only the minor difference that the absolute value is used to express T. For systems which include damping this identity is no longer true.

(2) The range of speeds for which T has been calculated is well above the resonance speed of about 120 rev/min. The results show that in this range the isolator is working well with values of T all considerably less than unity. Thus only a small proportion of the applied force is being transmitted to the ground or foundation. Perceptive readers will realize that the remainder is balanced by the inertia force due to motion of the mass.

PROBLEMS

(2.1) Using either Equation (2.2) or (2.3) and the first and second time derivatives, write a program to calculate the amplitude of motion and the system energy (Equation (2.5)) for given values of the system properties and given initial conditions (at time $t = 0$).

(2.2) Modify Example 2.5 to give the option of printing out the transmissibility in units of decibels (dB) of attenuation. This is a common unit and is particularly useful where isolators are used to attenuate sound radiation from a machine, because sound is universally measured in dB. The dB is a logarithmic unit which is appropriate where there is a wide range of values and where the measurand is related to human response. Most human sensations, including vision, hearing and sensation to vibration, are approximately logarithmic; that is to say, each time the stimulus is doubled the increase in sensation is a constant amount. Thus a logarithmic unit tends to measure sensation on a linear scale. In this problem the dB of attenuation may be defined as

$$dB = 20 \log_{10} T$$

and the program should be written accordingly. Include in the printout a caption indicating the units used for T.

(2.3) Write a program to compute the natural frequency of vertical vibrations of a prismatic body floating in a fluid of density ρ_f, with the prismatic axis vertical. The theory referring to Figure 2.5, is as follows:

The body is of length h (shown) and cross-sectional area A and floats in equilibrium, with the weight balanced by the Archimedean upthrust. This upthrust is equal to the weight of fluid displaced. A small depression of the body, x, as shown, increases the upthrust by an amount $\rho_f A x g$. This force is similar to the restoring force from a

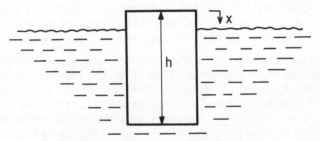

Figure 2.5 Prismatic body floating in a fluid

spring because it is proportional to displacement. The equivalent stiffness is

$$k = \rho_f A g$$

and so the natural frequency ω_n is given by the equations

$$\omega_n^2 = \frac{\rho_f A g}{m} \quad \text{or} \quad \omega_n^2 = \frac{\rho_f g}{\rho_b h} \tag{2.11}$$

where the mass of the body $m = \rho_b A h$. The program is to give the option of using either of the forms of Equation (2.11). Allow the usual options on units for natural frequency. How must the program be modified to solve the same problem for a non-prismatic body such as a ship?

In practice some of the fluid will move with the floating body as it vibrates and so an addition to the mass of the body must be made. This is called the virtual mass of the fluid. For most ships the virtual mass is of the same order as the mass of the ship itself. For vessels which have a high ratio of beam to draught, such as barges and landing craft, the virtual mass can be as high as three times the ship mass.

(2.4) The response curve for an accelerometer is obtained by solving Equation (2.9) for a ground acceleration of constant magnitude X and of varying frequency ω, as follows

$$\ddot{x}_0 = X \cos \omega t$$

The quantity which will be measured is the amplitude of y (Y). For an ideal accelerometer Y/X should be a constant, independent of frequency. This will obviously not be true at the resonance frequency. However the response curve, in terms of Y/X, will be nearly flat for quite a wide frequency band. Develop the necessary theory to obtain an expression for Y/X as a function of frequency.

Use the result to write a program, given the resonance frequency of the accelerometer and an allowed percentage deviation from the low frequency constant value for Y/X, will print out the upper frequency limit for operation of the accelerometer.

(2.5) Many pieces of machinery, mechanical instruments and electronic packages have to operate in environments where they may be subjected to impacts. The forces arising from these impacts are potentially capable of damaging these equipments and so they need protection. This usually takes the form of packaging material or isolators. In simple terms such systems may be modelled as in Figure 2.1 with a relative motion response given by Equation (2.2) or (2.3). The initial condition for determining the constants $(A$ and $B)$ or $(M$

and ϕ) is that $x = 0$ and $\dot{x} = v$, where v is the speed at impact. Derive the appropriate solution and use it to write a computer program which will determine the peak acceleration which the equipment will be subjected to and the peak relative displacement, given the mass of the equipment and the stiffness of the packaging. If the peak displacement exceeds a certain limit the packaging or isolator spring will bottom out and much larger accelerations will result. Arrange for the program to request a displacement limit and to print out a warning if it is exceeded.

Chapter 3

Simple systems with damping

Damping is the element, present in all real systems, which dissipates vibrational energy, usually as heat, and so attenuates the motion. It can take several forms depending on the physical mechanism involved. First, it can arise from the viscous forces due to shearing in fluids, in which case the damping force is proportional to velocity. Secondly, it may be caused by turbulent flows in fluids and the force is then proportional to velocity squared. Thirdly, it can arise from friction and the force is then nominally a constant but always acting to oppose the relative motion. This type of damping is often called Coulomb damping. Fourthly, its source can be within materials as a result of hysteresis in the stress–strain cycle. There can also be a mixture of these various types. In this book only the first type of damping, viscous damping, will be considered. This is for two reasons; first, because it leads most easily to analytical results, and secondly, because in most cases it is sufficiently representative of real damping to give accurate results.

ESSENTIAL THEORY

3.1 Equation of free motion

The shock absorber in motor vehicle suspensions is a typical viscous damping element. It generally consists of a piston which moves in a fluid-filled cylinder. The piston usually has a series of small holes in it through which the fluid is forced when the piston moves relative to the cylinder. Thus there is resistance to movement and this force is proportional to the relative velocity between the ends. The constant of proportionality, c, is called the damping coefficient. The simple system with damping included is now modelled as in Figure 3.1 where it will be seen that the conventional representation of a viscous damper is a piston in a cylinder.

Application of Newton's second law of motion to the mass in Figure 3.1, displaced by x as shown, gives the equation of motion

$$m\ddot{x} + c\dot{x} + kx = 0 \qquad (3.1)$$

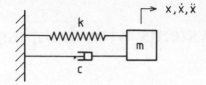

Figure 3.1 Model of the simple system with damping

If a trial solution $x = A e^{\lambda t}$ is substituted in Equation (3.1) this leads to an equation from which λ can be determined

$$m\lambda^2 + c\lambda + k = 0 \tag{3.2}$$

giving

$$\lambda = \frac{-c \pm \sqrt{c^2 - 4mk}}{2m} \tag{3.3}$$

The character of the solution clearly depends on whether c is greater or less than $2\sqrt{mk}$. Thus a critical damping coefficient is defined as

$$c_c = 2\sqrt{mk} = 2m\omega_n \tag{3.4}$$

where Equation (2.4) is used to define ω_n. However, ω_n is not now the natural frequency of vibration, as will be shown shortly. A non-dimensional damping ratio is defined by

$$\zeta = c/c_c \tag{3.5}$$

and Equation (3.3) becomes

$$\lambda = -\zeta\omega_n \pm \omega_n\sqrt{\zeta^2 - 1} \tag{3.6}$$

(i) High damping for which $\zeta > 1$

The quantity under the square root sign in Equation (3.6) is positive and substitution in the assumed form of solution leads to a sum of two decaying exponentials. The motion is so highly damped that no vibration occurs and this case will not therefore be considered further.

(ii) Low damping for which $\zeta < 1$

The quantity under the square root sign is now negative and so the complex number $i = \sqrt{-1}$ is introduced. Substitution in the assumed form of solution then gives

$$x = e^{-\zeta\omega_n t}(A\, e^{i\omega_n\sqrt{1-\zeta^2}t} + B\, e^{-i\omega_n\sqrt{1-\zeta^2}t})$$

This is developed by using de Moivre's theorem

$$e^{i\theta} = \cos\theta + i\sin\theta \tag{3.7}$$

leading to

$$x = e^{-\zeta\omega_n t}(M\cos(\omega_n\sqrt{1-\zeta^2}t) + N\sin(\omega_n\sqrt{1-\zeta^2}t)) \tag{3.8}$$

This solution shows a general sinusoidal oscillation at the damped natural frequency of $\omega_n\sqrt{1-\zeta^2}$ which decays according to the negative exponential factor $e^{-\zeta\omega_n t}$. The arbitrary constants M and N are determined from the initial conditions.

3.2 The logarithmic decrement

Equation (3.8) describes the decaying free oscillation which results from vibrational energy dissipation due to damping. This decay is often measured experimentally as a means of determining the amount of damping in a system. Figure 3.2 shows such a decay with the time origin chosen to make $N = 0$.

It is clear that damping can be described by the decay per cycle from X_1 to X_2 or from X_2 to X_3. A quantity called the logarithmic decrement (log. dec. for short, and symbol δ) is used for this purpose, defined by the equation

$$\delta = \log\left(\frac{X_1}{X_2}\right) \quad \left[\text{or} \quad \delta = \log\left(\frac{X_2}{X_3}\right)\right] \tag{3.9}$$

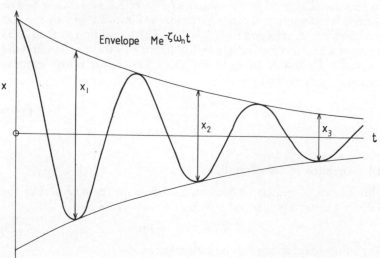

Figure 3.2 Record of vibration decaying as Equation (3.8)

where log is the natural logarithm (BASIC instruction LOG as defined in Chapter 1). Now

$$X_1 = 2M e^{-\zeta\omega_n t}$$

and

$$X_2 = 2M e^{-\zeta\omega_n(t+T)}$$

where T is the period of oscillation $2\pi/\omega_n\sqrt{1 - \zeta^2}$. Thus

$$\delta = \log(e^{\zeta\omega_n T}) = 2\pi\zeta/\sqrt{1 - \zeta^2} \tag{3.10}$$

Equations (3.4) and (3.10) define the relationships between three of the common forms of describing damping c, ζ and δ. When ζ is small (less than 0.1), Equation (3.10) is often simplified to

$$\delta \simeq 2\pi\zeta \tag{3.11}$$

3.3 Impulse response function

An arbitrary forcing function $F(t)$ applied to the mass in Figure 3.1 (as in Section 2.3) can be considered as a series of short impulses $F(t)\delta t$ and the total response may then be calculated as the corresponding sum of the responses to each individual impulse (see Section 3.6.2). This technique is only valid for systems obeying linear differential equations such as Equation (3.1). To calculate such a sum it is necessary to know the response to a unit impulse. Suppose that such an impulse is applied to the system of Figure 3.1 at time $t = 0$. The change in momentum is equal to the applied impulse (of value 1) and this gives an initial condition $\dot{x} = 1/m$. This condition is then used together with the other initial condition that $x = 0$ at $t = 0$ to find the constants M and N in Equation (3.8). The result is the impulse response function $h(t)$ given by

$$h(t) = \frac{e^{-\zeta\omega_n t}\sin(\omega_n\sqrt{1 - \zeta^2}\,t)}{m\omega_n\sqrt{1 - \zeta^2}} \tag{3.12}$$

3.4 Equation of forced motion

This section is similar to Section 2.3 but with the addition of the damping term. The general equation of motion is

$$m\ddot{x} + c\dot{x} + kx = F(t) \tag{3.13}$$

For sinusoidal forcing this equation becomes

$$m\ddot{x} + c\dot{x} + kx = F_0\cos\omega t \tag{3.14}$$

In the initial stages of motion there is a transient component given by Equation (3.8). However, this eventually decays and only the so-called particular integral of Equation (3.14) remains. This is the steady state solution (as defined previously in Chapter 2) and what follows is concerned exclusively with this part of the solution. It is most easily found by using complex numbers. Equation (3.14) may be regarded as the real part of a complex number differential equation

$$m\ddot{z} + c\dot{z} + kz = F_0\, e^{i\omega t} \tag{3.15}$$

where $z = x + iy$ and de Moivre's theorem, Equation (3.7), has again been used. A trial solution

$$z = X\, e^{i(\omega t - \phi)}$$

is substituted in Equation (3.15) with X defined as a real number. Under this assumption

$$x = X\cos(\omega t - \phi) \tag{3.16}$$

so that X is the amplitude of the motion and ϕ is the phase lag of the motion with respect to the applied force. The substitution leads by a few simple steps to the results

$$X = \frac{F_0/k}{\left[\left(1 - \left(\frac{\omega}{\omega_n}\right)^2\right)^2 + 4\zeta^2\left(\frac{\omega}{\omega_n}\right)^2\right]^{1/2}} \tag{3.17}$$

and

$$\tan\phi = \frac{2\zeta\left(\frac{\omega}{\omega_n}\right)}{\left[1 - \left(\frac{\omega}{\omega_n}\right)^2\right]} \tag{3.18}$$

A graph of Equation (3.17) is the classic resonance curve. This is shown for several values of ζ in Figure 3.3. It is plotted as the Dynamic Magnification Factor (DMF) against the frequency ratio (ω/ω_n). In Equation (3.17) the quantity F_0/k is the deflection of the mass under the action of a static force of magnitude F_0. Under the action of a dynamic force F_0 the deflection amplitude is X. Hence the DMF is given by

$$\text{DMF} = Xk/F_0 \tag{3.19}$$

The peak DMF occurs at a frequency of $\omega_n\sqrt{1 - 2\zeta^2}$ and this is called the resonance frequency. From Equation (3.17) it is seen that the phase lag is always 90° when $\omega/\omega_n = 1$ regardless of the damping.

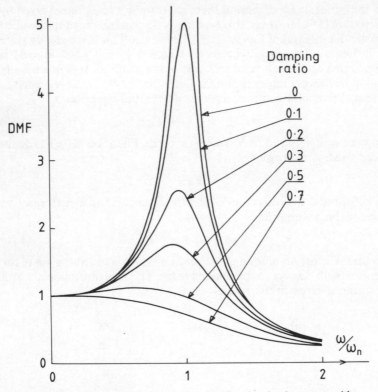

Figure 3.3 Frequency response function for the simple system with various damping ratios

This is called phase resonance. At this value of forcing frequency the DMF is $1/2\zeta$ and this is often measured experimentally as another means of evaluating the damping in a system.

A further method of measuring the damping is from the resonance curve and is known as the bandwidth method. Two points are marked on the curve which have a DMF of 0.707 times the peak DMF. These points define the frequency bandwidth and are often referred to as the 3 dB-down or half-power points. A decibel scale can be used for the DMF where

$$dB = 20 \log_{10}(DMF) \tag{3.20}$$

and it follows that a relative value of 0.707 or $1/\sqrt{2}$ corresponds to a 3 dB change. The difference between the values of ω/ω_n at the two points can be shown to be approximately equal to 2ζ when ζ is small.

The points also correspond approximately to frequencies at which the phase lags are 45° and 135°.

3.5 Fourier series analysis

A steady and continuous forcing of vibration may not necessarily be sinusoidal. It may be repetitive however, such as a square wave or a saw-tooth wave. If this is so then the forcing term can be expressed as a convergent series sum of sinusoids using the method of Fourier analysis. Because the system is linear the response may also be obtained as a sum of the responses to the individual sinusoids into which the force waveform has been decomposed. Each individual term in the response sum will then take the form of Equation (3.16). From these terms the response waveform can be constructed. The Fourier series for both force and response, which have an infinite number of terms, have to be truncated in order to make the calculation practicable. How this is done is a matter of judgment involving a number of factors including (1) the ratio of force repetition rate to system natural frequency, (2) the number of terms required in the series for the forcing function in order to obtain a reasonable approximation to the true response, and (3) the storage capacity of the computer used.

Suppose that the forcing function $F(t)$ is repetitive at intervals of T seconds and that the shape of its waveform is known. Then the Fourier series representation of this, with $\omega = 2\pi/T$, is

$$F(t) = A_0 + A_1 \cos(\omega t - \phi_1) + A_2 \cos(2\omega t - \phi_2) + \cdots \text{etc.}$$

(3.21)

or, since

$$\cos(n\omega t - \phi_n) = \cos\phi_n \cos n\omega t + \sin\phi_n \sin n\omega t$$

Equation (3.21) may be alternatively expressed

$$F(t) = A_0 + a_1 \cos\omega t + a_2 \cos 2\omega t + \cdots + b_1 \sin\omega t + b_2 \sin 2\omega t + \cdots$$

(3.22)

where $a_n = A_n \cos\phi_n$ and $b_n = A_n \sin\phi_n$ and thus

$$\begin{aligned} A_n &= (a_n^2 + n_n^2)^{1/2} \\ \tan\phi_n &= b_n/a_n \end{aligned}$$

(3.23)

It can easily be shown that A_0 is the mean value of $F(t)$, and this is often zero. The values for a_n and b_n are given by the integrals

$$a_n = \frac{2}{T} \int_{-T/2}^{+T/2} F(t) \cos n\omega t \, dt$$

$$b_n = \frac{2}{T} \int_{-T/2}^{+T/2} F(t) \sin n\omega t \, dt$$

(3.24)

If the origin for t can be chosen so that $F(t)$ is either symmetric or antisymmetric, then either all values of b or all values of a are zero. This will simplify calculations considerably.

The forcing function $F(t)$ will often be described, within the fundamental interval T, by a fixed number of values (p, taken here to be an odd number) at regular time intervals of $\Delta T = T/p$. In such a case the infinite series given by Equation (3.22) must be truncated to

$$F(t) = A_0 + \sum a_n \cos n\omega t + \sum b_n \sin n\omega t \qquad (3.25)$$

where the summations are taken from $n = 1$ to $n = (p - 1)/2$. A convenient simple evaluation of the Equations (3.24) can be obtained from

$$a_n = \frac{2}{p} \sum_{r=1}^{p} F(t_r) \cos n\omega t_r$$

$$b_n = \frac{2}{p} \sum_{r=1}^{p} F(t_r) \sin n\omega t_r$$

(3.26)

where the starting point for the summation with $r = 1$ is such that $t = -T/2$. More accurate evaluations can be obtained using the trapezium rule or Simpson's rule to perform the integrals (3.24) as numerical summations, see Further reading, p. 115, ref. 4, p. 122.

The continuous forcing function may also contain a random element such as in the excitation of tall buildings by wind. The calculation of the response in such cases is also based on Fourier analysis but is beyond the scope of this book. However, see Further reading, ref. 3, for details of the appropriate theory.

3.6   Transient forcing

The previous section has dealt with steady state vibration forcing. Engineers are also concerned occasionally with forcing in which the transient response is important, such as shock loading and abrupt changes in continuous loadings. If the forcing function is simple it may be possible to obtain an analytical solution to Equation (3.13). One technique is the method of Laplace transforms (in Further reading see ref. 3, Section 1.5). However, this is appropriate only where the Laplace transform of the forcing function has a well defined value. Numerical methods for calculating the response may be used where other methods fail and they are appropriate to this book. Two straightforward numerical methods are described below.

3.6.1 Step-by-step integration

In this method, a progressive calculation is made of the response quantities x, $\dot{x}$ and $\ddot{x}$ at discrete time intervals δt. It makes use of the Taylor expansions

$$x(t + \delta t) = x(t) + \delta t \dot{x}(t) + \frac{(\delta t)^2}{2} \ddot{x}(t) + \cdots \qquad (3.27)$$

$$\dot{x}(t + \delta t) = \dot{x}(t) + \delta t \ddot{x}(t) + \frac{(\delta t)^2}{2} \dddot{x}(t) + \cdots \qquad (3.28)$$

If δt is sufficiently small, then these equations may be truncated after two terms, hence ignoring terms in $(\delta t)^2$ and above. In procedures of this kind the step length should be, at most, one-tenth of the vibration period, for accurate work. The equation of motion (3.13) may be rearranged to give

$$\ddot{x}(t) = \frac{1}{m} \left(F(t) - c\dot{x}(t) - kx(t) \right) \qquad (3.29)$$

Generally, the initial values of x and $\dot{x}$ are known and are often zero. Hence the computational procedure would be as follows, denoting initial conditions with subscript 0 and subsequent values by subscripts $1, 2, 3, \ldots$, etc.

(1) Input x_0, $\dot{x}_0$, δt: set $t = 0$;
(2) Calculate $\ddot{x}_0$ from Equation (3.29);
(3) Calculate x_1, $\dot{x}_1$ from truncated Equations (3.27) and (3.28);
(4) Increment t by δt;
(5) Loop back to step 2 to calculate $\ddot{x}_1$;
(6) Repeat loop steps 2 to 5 as desired.

This procedure, which is of the very simplest kind, is rather inaccurate and in some cases may be unstable. However, it has the advantage of being easy to understand. It may be developed and improved leading, amongst others, to the rather more complicated central difference method, see Further reading, p. 115, ref. 3. A relatively simple improvement is given in ref. 5, p. 121.

3.6.2   Duhamel's integral method

This method is based on the use of the Impulse Response Function $h(t)$, as given by Equation (3.12) for the simple system with damping. The forcing function $F(t)$ is regarded as a series of impulses of magnitude $F(t)\delta t$ occurring at time t. The response at some later time τ caused by one such impulse is

$$\delta x(\tau) = F(t)\delta t\, h(\tau - t)$$

If the effect of all such impulses is summed (integrated) then the total response is given by the Duhamel or superposition integral

$$x(\tau) = \int_{-\infty}^{\tau} F(t)h(\tau - t)\,\mathrm{d}t \qquad (3.30)$$

If this integral is to be calculated step by step in a computer program, and if $F(t)$ is zero for $t < 0$, then it can be expressed as the sum

$$x(n\delta t) = \sum_{t=0}^{t=n\delta t} F(t)\delta t\, h(n\delta t - t) \qquad (3.31)$$

This sum gives the displacement value at only one instant of time and therefore has to be repetitively calculated if the displacement waveform is to be found.

3.7   Rotating machinery

Any machine containing rotating parts is a source of vibration and, indeed, such machines are one of the commonest sources of vibration problems in engineering. In this section two important topics are covered; first, the quality of vibration forcing due to lack of balance, and secondly, the problem of shaft whirl.

3.7.1   Vibration forcing caused by lack of balance

A machine of total mass m is mounted vertically on a spring and damper with properties k and c respectively. Its movement is described by the coordinate x as in Figure 3.1. It contains a light horizontal shaft rotating at rate ω and having a disc of mass m' mounted on it. If the disc is unbalanced so that its centre of gravity is at a distance e (eccentricity) from the axis of rotation then the vertical motion y of mass m' may be written

$$y = x + e \cos \omega t$$

The equation for vertical motion of the whole system may then be written

$$(m - m')\ddot{x} + m'(\ddot{x} - e\omega^2 \cos \omega t) + c\dot{x} + kx = 0$$

or

$$m\ddot{x} + c\dot{x} + kx = m'e\omega^2 \cos \omega t \tag{3.32}$$

This is exactly the same as Equation (3.14) except that F_0 is replaced by $m'e\omega^2$. The solution is obtained similarly and the displacement amplitude is given by

$$X = \frac{\dfrac{em'}{m}\left(\dfrac{\omega}{\omega_n}\right)^2}{\left[\left(1 - \left(\dfrac{\omega}{\omega_n}\right)^2\right)^2 + 4\zeta^2 \left(\dfrac{\omega}{\omega_n}\right)^2\right]^{1/2}} \tag{3.33}$$

The equation for phase angle ϕ is identical to Equation (3.18).

3.7.2 Shaft whirl

If a flexible shaft rotates at a speed equal to its natural frequency of lateral bending vibration then the lateral deflection may become very large. This phenomenon may be easily demonstrated and is known as 'whirl'. It is potentially dangerous and destructive and must be avoided. In design, the whirl speed may be determined by finding the natural frequency of static lateral vibration of the shaft (see Chapter 6).

For an undamped light shaft with a disc of mass m' mounted on it, as in the previous section, the analysis is straightforward. Assume that the lateral stiffness of the shaft is k' at the point of attachment of the disc. Referring to Figure 3.4, 0 is on the straight dotted line joining the bearing centres and is at the geometric centre of the disc when it is stationary. A is the deflected position of the geometric centre when the disc is rotating with $0A = r$. G is the centre of gravity of the disc with $AG = e$. Then the outward centrifugal force is due to $(r + e)$ but the

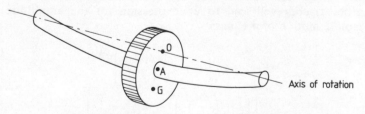

Figure 3.4 A simple shaft-disc system in whirl

restoring force is due to r alone. For equilibrium

$$m'(r + e)\omega^2 - k'r = 0$$

and this gives for the elastic deflection

$$r = \frac{e\left(\dfrac{\omega}{\omega_n}\right)^2}{\left[1 - \left(\dfrac{\omega}{\omega_n}\right)^2\right]} \quad \text{with} \quad \omega_n^2 = \frac{k'}{m'} \tag{3.34}$$

Equation (3.34) shows the elastic divergence of lateral deflection r which has been described above. The inclusion of damping complicates the analysis (see ref. 3, Further reading, p. 115) and results in limited motion at the whirl speed.

3.8   Earth motion

The analysis of earth motion for a damped simple system is similar to that of Section 2.4 for the undamped system except that the equation for relative motion, y is given by

$$m\ddot{y} + c\dot{y} + ky = -m\ddot{x}_0 \tag{3.35}$$

If $x_0 = X_0 \cos \omega t$ then F_0 in Equation (3.14) is replaced by $m\omega^2 X_0$ and substitution in Equation (3.17) gives for the amplitude of relative motion

$$Y = \frac{X_0 \left(\dfrac{\omega}{\omega_n}\right)^2}{\left[\left(1 - \left(\dfrac{\omega}{\omega_n}\right)^2\right)^2 + 4\zeta^2 \left(\dfrac{\omega}{\omega_n}\right)^2\right]^{1/2}} \tag{3.36}$$

Note the marked similarity between this and Equation (3.33). Some further algebra will lead to an expression for the amplitude of absolute motion of the mass

$$X = \frac{X_0 \left[1 + 4\zeta^2 \left(\dfrac{\omega}{\omega_n}\right)^2\right]^{1/2}}{\left[\left(1 - \left(\dfrac{\omega}{\omega_n}\right)^2\right)^2 + 4\zeta^2 \left(\dfrac{\omega}{\omega_n}\right)^2\right]^{1/2}} \tag{3.37}$$

WORKED EXAMPLES

Example 3.1 Quadratic equations: damping ratio and natural frequency

If the mass, stiffness and damping coefficient are known for a simple system then the natural frequency and damping ratio may easily be determined from the solution to the quadratic Equation (3.2). The following simple program calculates the roots of the quadratic equation in its conventional general form

$$ax^2 + bx + c = 0$$

The output indicates whether the roots are real or complex and this corresponds to the two cases (1) $\zeta > 1$, and (2) $\zeta < 1$. The sample run is for a machine for which $m = 220\,kg$, $c = 2775\,kg/s$ and $k = 35\,kN/m$.

```
10   REM   EX3POINT1
20   PRINT "SOLUTION TO THE QUADRATIC EQUATION AX^2+BX+C=0"
30   PRINT "------------------------------------------------"
40   PRINT "INPUT A,B,C ?"
50   INPUT A,B,C
60   PRINT
70   D = B * B - 4 * A * C
80   IF D < 0 GOTO 170
90   PRINT "REAL ROOTS"
100   PRINT "----------"
110  E =  SQR (D)
120  X1 = ( - B + E) / (2 * A)
130  X2 = ( - B - E) / (2 * A)
140   PRINT "X1 = ";X1
150   PRINT "X2 = ";X2
160   GOTO 320
170   PRINT "COMPLEX ROOTS"
180   PRINT "-------------"
190  E =  SQR ( - D) / (2 * A)
200  F =  - B / (2 * A)
210   PRINT "X1 = ";F;" + (J)*";E
220   PRINT "X2 = ";F;" - (J)*";E
230   PRINT
240   PRINT "VIBRATION QUANTITIES"
250   PRINT "--------------------"
260  Z = B / (2 *  SQR (A * C))
270  H = 2 * 3.14159
280   PRINT "UNDAMPED NATURAL FREQUENCY = ";( - F / (H * Z));" HZ"
290   PRINT "DAMPED NATURAL FREQUENCY = ";E / H;" HZ"
300   PRINT "DAMPING RATIO = ";Z
310   PRINT
320   PRINT "ANOTHER RUN?"
330   GET C$
340   IF C$ = "Y" THEN 30
350   PRINT "-------------------------------------------------"
360   STOP

JRUN
SOLUTION TO THE QUADRATIC EQUATION AX^2+BX+C=0
------------------------------------------------
INPUT A,B,C ?
?220,2775,35000
```

```
COMPLEX ROOTS
--------------
X1 = -6.30681818 + (J)*10.9231385
X2 = -6.30681818 - (J)*10.9231385

VIBRATION QUANTITIES
--------------------
UNDAMPED NATURAL FREQUENCY = 2.0074428 HZ
DAMPED NATURAL FREQUENCY = 1.73847295 HZ
DAMPING RATIO = .500020292

ANOTHER RUN?
-----------------------------------------

BREAK IN 360
```

Program note

(1) Lines 230 to 310 give results for the vibrational Equation (3.2) in the usual format of natural frequencies and damping ratio. If these lines are deleted the program provides a general solution to quadratic equations. For real roots no vibration quantities have any meaning and so after printing the roots the program stops.

Example 3.2   Polar response diagrams

An experimental technique often used to analyse vibrations in mechanical components is to measure the amplitude of forced vibration (see Equation (3.17)), and the corresponding phase angle, ϕ (see Equation (3.18)) for a range of discrete frequencies, ω_1, ω_2, ..., etc. The information gained is then plotted on polar graph paper to form a Polar or Vector Response Diagram. From this graph the undamped natural frequency can be determined by interpolation for the point at which $\phi = 90°$ and the damping can be found by a development of the bandwidth method (see Section 3.4) provided that the damping ratio is small.

This program calculates the values of X and ϕ using Equations (3.17) and (3.18) for a range of frequencies. It also converts these values to rectangular coordinates x and y so that the polar diagram can be plotted on ordinary graph paper. If the reader is prepared to venture into computer graphics the program may easily be extended to display the polar diagram.

In the sample calculation the system properties are the same as those used in the previous example. The force magnitude is given as 100 N and the range of frequencies for which calculations are made is from 5 to 20 rad/s in steps of 1 rad/s. Figure 3.5 is a plot of the polar response obtained.

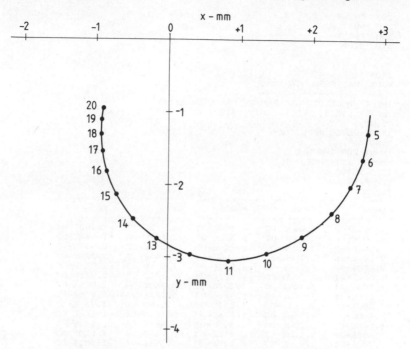

Figure 3.5 Polar response diagram obtained in Example 3.2

```
10   REM EX3POINT2
20   PRINT "POLAR RESPONSE DIAGRAMS"
30   PRINT "------------------------"
40   PRINT "INPUT RANGE OF FREQUENCIES AND STEP LENGTH"
50   PRINT "(LOWER LIMIT, UPPER LIMIT, STEP) ?"
60   INPUT I,J,L
70   PRINT "INPUT FORCE AMPLITUDE ?"
80   INPUT F
90   PRINT "INPUT FOR THE SYSTEM"
100  PRINT "(MASS, STIFFNESS) ?"
110  INPUT M,K
120  PRINT "(DAMPING COEFFICIENT) ?"
130  INPUT C
140  N = SQR (K / M)
150  Z = C / (2 * SQR (M * K))
160  S = F / K
170  PRINT "-----------------------------------------------------------"
180  PRINT "FREQ.","MAGN.","PHASE","X","Y"
190  PRINT "-----------------------------------------------------------"
200  FOR W = I TO J STEP L
210  R = W / N
220  A = 2 * Z * R
230  B = 1 - (R * R)
240  X = S / SQR ((B * B) + (A * A))
250  P = ATN (A / B)
260  IF P > 0 THEN 280
270  P = P + 3.14159265
```

```
280   PRINT W,X,P,(X * COS (P)), - (X * SIN (P))
290   NEXT W
300   PRINT "---------------------------------------------------------"
310   STOP


JRUN
POLAR RESPONSE DIAGRAMS
------------------------
INPUT RANGE OF FREQUENCIES AND STEP LENGTH
(LOWER LIMIT, UPPER LIMIT, STEP) ?
?5,20,1
INPUT FORCE AMPLITUDE ?
?100
INPUT FOR THE SYSTEM
(MASS, STIFFNESS) ?
?220,35000
(DAMPING COEFFICIENT) ?
?2775
```

FREQ.	MAGN.	PHASE	X	Y
5.000E+00	3.067E-03	4.396E-01	2.775E-03	-1.305E-03
6.000E+00	3.145E-03	5.512E-01	2.679E-03	-1.647E-03
7.000E+00	3.220E-03	6.759E-01	2.512E-03	-2.015E-03
8.000E+00	3.278E-03	8.150E-01	2.248E-03	-2.385E-03
9.000E+00	3.298E-03	9.682E-01	1.869E-03	-2.717E-03
1.000E+01	3.263E-03	1.132E+00	1.384E-03	-2.955E-03
1.100E+01	3.159E-03	1.302E+00	8.363E-04	-3.046E-03
1.200E+01	2.988E-03	1.471E+00	2.964E-04	-2.973E-03
1.300E+01	2.766E-03	1.631E+00	-1.669E-04	-2.761E-03
1.400E+01	2.519E-03	1.776E+00	-5.154E-04	-2.466E-03
1.500E+01	2.268E-03	1.905E+00	-7.463E-04	-2.142E-03
1.600E+01	2.030E-03	2.018E+00	-8.788E-04	-1.830E-03
1.700E+01	1.813E-03	2.115E+00	-9.394E-04	-1.550E-03
1.800E+01	1.619E-03	2.198E+00	-9.519E-04	-1.310E-03
1.900E+01	1.450E-03	2.270E+00	-9.345E-04	-1.109E-03
2.000E+01	1.303E-03	2.333E+00	-8.999E-04	-9.423E-04

Program notes

(1) The function ATN (line 250) returns a value for $\phi(P)$ which lies between $-\pi/2$ and $\pi/2$ so that when P is negative it has to be corrected by adding π to obtain a value for ϕ in the range 0 to π.

(2) The printout of results has been modified to contain them within a width which can be reproduced in this book. This modification is not shown in the given program because the formation of PRINT statements has wide variations between different computers. The PRINT statement in the program gives standard tabulation.

(3) The program may easily be expanded to accommodate options to repeat the calculation by looping back for (1) different values of damping coefficient, (2) different mass and stiffness values, (3) different force amplitudes, and (4) different ranges of frequencies.

(4) By interpolation on Figure 3.5 it is possible to determine the undamped natural frequency at $\phi = 90°$ ($\omega_n \simeq 12.6$ rad/s). From the

printed table the peak motion amplitude is about 3.3 mm at a value of frequency between 8 and 9 rad/s.

(5) The bandwidth method of determining damping may not be used here because the damping value is too high. If the program is re-RUN for $c = 277.5$ (one-tenth of the previously used damping and giving $\zeta = 0.05$) the bandwidth method will give an accurate value for damping ratio. It will also be seen that for small damping the polar response diagram closely approximates a circle. Indeed, the theory for hysteretic damping shows that the diagram is a circle passing through the origin with a diameter coincident with the negative y-axis.

Example 3.3   Transient forcing: step-by-step integration

This example follows the theory given in Section 3.6.1 for the machine system which has been used to illustrate all the examples in this chapter. The transient force applied to the machine is an impulsive blow which has the triangular waveform shown in Figure 3.6. This starts at time $t = 0$ and builds up to a peak of 1000 N at $t = 1$ after which the force is zero. The displacement response is required for three seconds from $t = 0$.

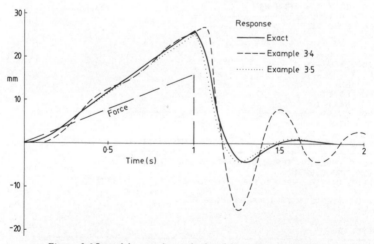

Figure 3.6 Impulsive motion calculated by various methods

In this program the step length has been chosen to be such that there are about ten steps per cycle of vibratory motion. Any less than this leads to unacceptable inaccuracies. A shorter time interval will generally lead to increased accuracy except that error due to accumulated rounding off can be a problem if there is a very great number of steps in the integration process.

This program has not been written in a very general form: The total time for integration has been written into instruction line 110 and the forcing function has been defined by lines 210 to 240 inclusive. The program can be generalized, without difficulty, but this makes it a good deal longer.

```
10   REM EX3POINT3
20   PRINT "TRANSIENT FORCING: STEP BY STEP INTEGRATION"
30   PRINT "--------------------------------------------------"
40   PRINT "INPUT (MASS, STIFFNESS, DAMPING COEFFICIENT)"
50   INPUT M,K,C
60   PRINT "INPUT INITIAL VALUES FOR (DISPLACEMENT, VELOCITY)"
70   INPUT D,V
80 T = 0
90   PRINT "INPUT STEP LENGTH IN SECONDS ?"
100  INPUT H
110  J =   INT (( INT (3 / H) + 1) / 5)
120  PRINT "--------"
130  PRINT "RESULTS"
140  PRINT "--------------------------------------------------------"
150  PRINT "TIME:  DISPLACEMENTS FOR FIVE SUCCESSIVE INTERVALS"
160  PRINT "--------------------------------------------------------"
170  FOR I = 0 TO J
180  PRINT T;"  ";
190  FOR L = 1 TO 5
200  PRINT D;"  ";
210  IF T > 1 THEN 240
220 F = 1000 * T
230  GOTO 250
240 F = 0
250 A = (F - C * V - K * D) / M
260 D = D + H * V
270 V = V + H * A
280 T = T + H
290  NEXT L
300  PRINT
310  NEXT I
320  PRINT "--------------------------------------------------------"
330  STOP
JRUN
TRANSIENT FORCING: STEP BY STEP INTEGRATION
------------------------------------------------
INPUT (MASS, STIFFNESS, DAMPING COEFFICIENT)
?220,35000,2775
INPUT INITIAL VALUES FOR (DISPLACEMENT, VELOCITY)
?0,0
INPUT STEP LENGTH IN SECONDS ?
?.05
--------
RESULTS
```

TIME:	DISPLACEMENTS FOR FIVE SUCCESSIVE INTERVALS				
0.00	0.00000	0.00000	0.00000	0.00056	0.00191
0.25	0.00389	0.00613	0.00825	0.01000	0.01134
0.50	0.01241	0.01340	0.01451	0.01584	0.01738
0.75	0.01903	0.02068	0.02224	0.02368	0.02503
1.00	0.02633	0.02765	0.01766	0.00298	-0.00946
1.25	-0.01525	-0.01362	-0.00695	0.00092	0.00660
1.50	0.00833	0.00634	0.00229	-0.00172	-0.00411

1.75	-0.00431	-0.00275	-0.00045	0.00148	0.00238
2.00	0.00212	0.00108	-0.00014	-0.00103	-0.00130
2.25	-0.00098	-0.00035	0.00027	0.00064	0.00067
2.50	0.00042	0.00007	-0.00023	-0.00037	-0.00033
2.75	-0.00016	0.00002	0.00016	0.00020	0.00015
2.99	0.00005	-0.00004	-0.00010	-0.00010	-0.00006

BREAK IN 330

Program notes

(1) The main loop for the integration process is controlled by I (line 170). There is a nested loop within this, controlled by L (line 190), which is designed to print out five consecutive results on one line so as to economize on paper. The time value is printed correctly at the start of the line for the first displacement value. Because the program prints out results in batches of five values the process will overshoot the set period of integration, in this case by four values. The program could be modified to prevent this but it would then become unnecessarily complicated. The printout has again been modified as in Example 3.2.

(2) The exact solution to Equation (3.13) for this problem has been derived as a check. The results from the program and the corresponding exact response are plotted on Figure 3.6. Comparison is not very good, particularly after the forcing has ceased (after $t = 1$). The reader is encouraged to try this program with different step lengths to see the effect that this has on accuracy. The plotting of such results is rather a tedious process and this is one case where graphics can save a great deal of time.

Example 3.4   Transient forcing: Duhamel's integral

In this example the problem of Example 3.3 is tackled by the Duhamel's integral method, described in Section 3.6.2. For this calculation the impulse response function needs to be known and for this problem the program uses the appropriate function given by Equation (3.12). This is first calculated at the appointed intervals and stored in a suitable array (H). This array can be terminated at the point where the function is judged to have died away sufficiently although in this case it is calculated for a full three seconds for programming convenience. The force function is also calculated and placed in an array (F) within the same program loop starting at line 130. These two arrays are then used to calculate the response using Equation (3.31) and this is printed out in the same way as for the previous example. The program assumes that the machine is initially at rest at time $t = 0$.

```
10   REM EX3POINT4
20   PRINT "TRANSIENT FORCING: DUHAMEL'S INTEGRAL"
30   PRINT "-------------------------------------"
40   PRINT "INPUT (MASS, STIFFNESS, DAMPING COEFFICIENT)"
50   INPUT M,K,C
60   PRINT "INPUT STEP LENGTH IN SECONDS ?"
70   INPUT H
80   J = INT (( INT (3 / H) + 1) / 5)
90   DIM H(100),D(100),F(100)
100  Z = C / (2 * SQR (M * K))
110  F = SQR ((K * (1 - Z * Z)) / M)
120  T = 0
130  FOR I = 0 TO (5 * J)
140  D(I) = 0
150  H(I) = EXP ( - Z * T * SQR (K / M)) * SIN (F * T) / (F * M)
160  IF T > 1 THEN 190
170  F(I) = 1000 * T
180  GOTO 200
190  F(I) = 0
200  T = T + H
210  NEXT I
220  FOR I = 0 TO (5 * J)
230  FOR L = 0 TO I
240  D(I) = D(I) + F(L) * H(I - L) * H
250  NEXT L
260  NEXT I
270  T = 0
280  PRINT "--------"
290  PRINT "RESULTS"
300  PRINT "----------------------------------------------------"
310  PRINT "TIME:  DISPLACEMENTS FOR FIVE SUCCESSIVE INTERVALS"
320  PRINT "----------------------------------------------------"
330  FOR I = 0 TO J
340  PRINT T;"  ";
350  FOR L = 0 TO 4
360  PRINT D((5 * I) + L);"  ";
370  T = T + H
380  NEXT L
390  PRINT
400  NEXT I
410  PRINT "----------------------------------------------------"
420  STOP
JRUN
TRANSIENT FORCING: DUHAMEL'S INTEGRAL
-------------------------------------
INPUT (MASS, STIFFNESS, DAMPING COEFFICIENT)
?220,35000,2775
INPUT STEP LENGTH IN SECONDS ?
?.05
--------
RESULTS
----------------------------------------------------
```

TIME:	DISPLACEMENTS FOR FIVE SUCCESSIVE INTERVALS				
0.00	0.00000	0.00000	0.00039	0.00127	0.00256
0.25	0.00409	0.00571	0.00730	0.00883	0.01027
0.50	0.01165	0.01300	0.01435	0.01570	0.01706
0.75	0.01843	0.01981	0.02120	0.02259	0.02397
1.00	0.02536	0.01886	0.01001	0.00245	-0.00227
1.25	-0.00414	-0.00395	-0.00272	-0.00129	-0.00016
1.50	0.00048	0.00069	0.00060	0.00038	0.00015
1.75	0.00000	-0.00009	-0.00011	-0.00009	-0.00005
2.00	-0.00001	0.00000	0.00001	0.00001	0.00001
2.25	0.00000	0.00000	0.00000	0.00000	0.00000
2.50	0.00000	0.00000	0.00000	0.00000	0.00000
2.75	0.00000	0.00000	0.00000	0.00000	0.00000
2.99	0.00000	0.00000	0.00000	0.00000	0.00000

```
----------------------------------------------------

BREAK IN 420
```

Program notes

(1) The arrays H, D and F have been dimensioned to contain a maximum of 100 points. Only 65 have been used in the sample run. The capacity can easily be increased, if required, by increasing the dimensions in line 90.

(2) Lines 220 to 260 inclusive constitute a pair of nested loops which calculate the displacement response values using Equation (3.31).

(3) The results have again been compressed in format for reproduction here. They have also been plotted on Figure 3.6 for comparison with the exact solution. This comparison is generally much better than for Example 3.3 and has the same step length. It will be found however that this program takes considerably longer to run than Example 3.3 because of the extensive calculations that are required. Once again the interested reader can try shorter step lengths to assess the effect that this has on the accuracy of the calculation. If this is attempted line 90 may need to be changed.

PROBLEMS

(3.1) Write a program to calculate the values of logarithmic decrement from a series of experimentally measured peak values from a free decay curve. Note that if the amplitude ratio is measured every n cycles, Equation (3.9) has to be extended to give

$$\delta = \frac{1}{n} \log \left(\frac{X_1}{X_{n+1}} \right)$$

Write the program to solve the following problem.

In a vibration test on the blade of a helicopter a free decay is measured and registered on a chart recorder. From the record the envelope amplitude is measured every four cycles and the results in mm are

108, 74, 51, 35, 24, 16

Show that these figures are consistent with viscous damping and derive the logarithmic decrement.

(3.2) Write a program to calculate the DMF as a function of sinusoidal forcing frequency for the case of a machine with rotating unbalance making use of Equation (3.33). For suitable values of damping ratio use the results to plot curves on the same basis as Figure 3.3 and comment upon the differences. Use the program also to solve the following special case.

A machine tool of total mass 850 kg contains within it an out-of-balance rotating shaft with a product of mass and eccentricity of 0.2 kgm. The whole machine is spring mounted and without damping. The force transmitted to the foundation has a magnitude of one-fifth of that of the exciting force when the shaft is rotating at 3000 rev/min. Use this information to find the stiffness of the spring mount.

Use the program to determine whether the force transmitted to ground will exceed 10 kN in the shaft speed range 1000 rev/min to 3000 rev/min and, if so, for what part of the range.

(3.3) Write a program to calculate both the relative and absolute response of a simple system with damping when subjected to sinusoidal earth motion making use of Equations (3.36) and (3.37). The program should obtain results for a range of values of (ω/ω_n) and is to be used to solve the following problem.

The springs of a vehicle trailer compress 120 mm under the weight of the trailer and its load and the shock absorbers are such that the damping ratio for vertical vibrations is 0.3.

The vehicle and its trailer pass, at various speeds, over a road with a surface profile approximated by a sine-wave of peak-to-peak amplitude 80 mm and wavelength 10 m. Plot a graph of the peak-to-peak absolute motion of the trailer as a function of road speed between 0 and 100 km/h.

(3.4) An accelerometer is modelled as a simple system with damping subject to earth motion. The earth motion to be measured contains a mixture of frequencies below the resonance frequency of the accelerometer. Each component of frequency ω will be delayed in the system response (and thus in the measurement) by a time, T, which depends upon both the phase lag, ϕ (Equation (3.18)), and the frequency, as follows

$$T = \phi/\omega$$

If the waveform of the acceleration is to be preserved in the measurement, each Fourier component should be subject to the same time delay. Write a computer program to calculate T in the frequency range $0 < \omega < \omega_n$ for a series of fixed values for ζ, the damping ratio. Plot these curves and hence derive the optimum value for the damping ratio to achieve constancy of T in the given frequency range.

(3.5) Use Equations (3.25) and (3.26) to construct a 21-point Fourier series for the square wave forcing function $F(t)$ for which

$$F(t) = +1 \quad \text{for} \quad -0.5 \leqslant t < 0.5$$
$$F(t) = -1 \quad \text{for} \quad -1 \leqslant t < -0.5 \quad \text{and} \quad 0.5 \leqslant t \leqslant 1$$

The first part of the program is to calculate the coefficients a_n from Equation (3.26). Note that coefficients b_n are all zero due to symmetry of the waveform about $t = 0$. The second part of the program is to reconstruct the waveform using Equation (3.25) to check the accuracy of the process. If the check is not deemed to be satisfactory repeat the computation for a 41-point series.

Chapter 4

Two degree of freedom systems

Although many engineering systems can be modelled satisfactorily by a simple system with one degree of freedom there are also many others for which this is not true. It is often necessary to take account of several degrees of freedom if the motion and component stresses are to be predicted sufficiently accurately for design purposes. An example is the motor vehicle mounted on a suspension spring at each corner. This system effectively has three degrees of freedom and three corresponding modes of vibration, namely pitch, roll and vertical motion. The analysis of such systems introduces several new aspects of vibration theory which have not been met in the earlier chapters. This chapter introduces these new concepts for systems with two degrees of freedom. For such systems analysis is possible by relatively simple means and thus a clear understanding may be gained of the new concepts. There are also a few special applications of two degree of freedom (2-DOF) analysis such as tuned vibration absorbers. The analysis of systems with many degrees of freedom (n-DOF) introduces no further new principles but the solution of problems becomes complicated and leads to the use of matrix methods. These are described in Chapter 5. It is important to appreciate however that the 2-DOF system is only a special case of n-DOF systems. The underlying principles are identical for all n-DOF systems even though it is convenient in this book to describe the methods of solution in two chapters.

ESSENTIAL THEORY

4.1 The equation of free motion without damping

The theory for 2-DOF systems may be described in a completely general way, with different algebraic symbols for each component. However, to appreciate the physical nature of what is happening, it is better to be more restrictive and to solve a specific problem. Thus, in what follows, each component property is numerically related to either a mass value, m or to a stiffness value, k. Figure 4.1 shows the

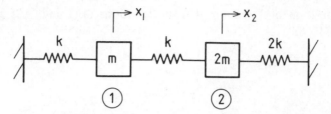

Figure 4.1 Model of a typical 2-DOF system used an an example

specific system which will be analysed here. The natural frequencies of this system may be numerically related to a fictitious frequency $\sqrt{k/m}$. The system has deliberately been made asymmetric and it is assumed that gravity does not act in any significant way. The procedure which will be followed is typical of that to be used for any system of this general kind.

The two independent degrees of freedom for this system are the displacements x_1 and x_2 of the masses. To obtain the equations of motion for the system the dynamic equilibrium of each mass is evaluated. This makes use of d'Alembert's principle for including the inertia force and the concept of a Free Body Diagram (FBD) to include all the forces which act at each mass position. The FBD for mass 1 is shown in Figure 4.2 and the corresponding equation of motion is

$$m\ddot{x}_1 + kx_1 - k(x_2 - x_1) = 0 \tag{4.1}$$

A similar FBD for mass 2 leads to the second equation of motion

$$2m\ddot{x}_2 + k(x_2 - x_1) + 2kx_2 = 0 \tag{4.2}$$

Thus the system is described by two simultaneous linear differential equations of motion. These may be solved by a trial solution which assumes that when the system vibrates harmonically both masses vibrate at the same frequency and in phase, as follows

$$\begin{aligned} x_1 &= X_1 \cos \omega t \\ x_2 &= X_2 \cos \omega t \end{aligned} \tag{4.3}$$

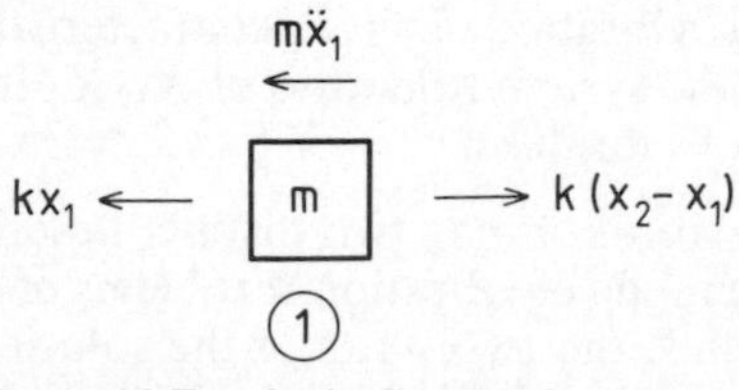

$$m\ddot{x}_1$$
$$\longleftarrow$$

$$kx_1 \longleftarrow \quad \boxed{m} \quad \longrightarrow k(x_2 - x_1)$$

①

Figure 4.2 Free body diagram for mass 1

Substitution of Equation (4.3) in Equations (4.1) and (4.2) gives, respectively

$$\frac{X_1}{X_2} = \frac{k}{2k - m\omega^2} = \frac{1}{2 - \phi} \qquad (4.4)$$

$$\frac{X_1}{X_2} = \frac{3k - 2m\omega^2}{k} = 3 - 2\phi \qquad (4.5)$$

$\phi = m\omega^2/k$ is a non-dimensional frequency parameter. Equations (4.4) and (4.5) lead to the characteristic or frequency equation

$$(3 - 2\phi)(2 - \phi) = 1$$
or $\qquad\qquad 2\phi^2 - 7\phi + 5 = 0 \qquad (4.6)$

This equation, by simple factorization, has two solutions

(1) $\phi = 1$ and, from Equation (4.5), $X_1/X_2 = 1$.
From these values the physical description of this vibration may be deduced. First, the vibration has a natural frequency which is equal to the frequency $\sqrt{k/m}$. Secondly, masses 1 and 2 move together and by the same amount. Thus the middle spring remains unstretched and, as far as this mode of vibration is concerned, it could be removed because it takes no active part in the motion. This view of things explains immediately why the natural frequency is $\sqrt{k/m}$. However, the result is specific to this chosen case and it is not generally true that the middle spring may be disregarded in order to find one of the modes of vibration.

(2) $\phi = 2.5$ and, from Equation (4.5), $X_1/X_2 = -2$.
This vibration has a higher natural frequency of $1.581\sqrt{k/m}$. The masses move in opposite directions ($-$ sign), and mass 1 (the lighter) moves twice as far as mass 2.

Some general comments about these solutions are appropriate at this point:

(1) 2-DOF systems have two natural frequencies. It can easily be seen that an n-DOF system will have n natural frequencies.

(2) Each natural vibration has a characteristic distribution of deflection within the system (the ratio of X_1/X_2 in this case). This is called the mode of vibration.

(3) Each natural vibration has two distinct descriptors; the natural frequency and the mode of vibration. Problems of this kind are called eigenvalue problems, the two parts of the solution being called the eigenvalue (frequency) and the eigenvector (mode). The eigenvalue

problem also occurs in other branches of engineering design, for example, in buckling theory.

(4) There is no solution for absolute motion, only for relative motion. Thus X_1 remains unknown or arbitrary, in any particular natural vibration, but X_1/X_2 is defined. This correlates with the intuitive knowledge that systems can vibrate at different amplitudes, depending on the initial conditions.

(5) The solution of lowest natural frequency ($\omega_1 = \sqrt{k/m}$) is often termed the fundamental. Higher solutions are often termed harmonics though in the strict musical sense this is an inept description. Thus in the example the second solution is for the first harmonic ($\omega_2 = 1.581\sqrt{k/m}$). It is perhaps better to call these the first and second natural frequencies with corresponding first and second natural modes of vibration.

(6) The complete solution to the equations of motion is a combination of the two modes of vibration each of which will depend for its amplitude on the initial conditions. Generally, motion in both modes will occur simultaneously as a linear superposition. If the motion of one of the masses is determined it will be found that this superposition will result in beating (see Example 4.3, p. 69). The general form of the resulting motion will be

$$x_1 = A_1 \cos(\omega_1 t + \phi_1) + B_1 \cos(\omega_2 t + \phi_2)$$
$$x_2 = A_2 \cos(\omega_1 t + \phi_1) + B_2 \cos(\omega_2 t + \phi_2) \tag{4.7}$$

where the ratios of A_1/A_2 and B_1/B_2 are defined by the modal ratios X_1/X_2 for each mode (in this case $+1$ and -2 respectively).

(7) A relatively simple model of this sytem can be constructed by the reader and may help with a physical appreciation of its vibration properties. The masses are modelled by a large nut for mass 1, two large nuts for mass 2. The springs are modelled by thin elastic bands about 150 mm long and attached to the nuts with a loop knot. If this sytem is stretched horizontally between two points about 0.8 m apart it can then be set into motion (vertical is easiest) by pulling the masses to one side of the equilibrium position and releasing them. Different relative initial displacements will produce different combinations of the two modes of vibration.

4.2   Coupling

Equation (4.1) is obtained by considering the motion of mass 1. However, the equation includes a term in x_2 as a result of the spring which joins the two masses. Thus the equation contains a stiffness or static coupling term. Indeed. Equations (4.1) and (4.2) can be written as a single matrix equation

$$\mathbf{M\ddot{x}} + \mathbf{Kx} = \mathbf{0} \tag{4.8}$$

or, in full

$$\begin{bmatrix} m & 0 \\ 0 & 2m \end{bmatrix} \begin{Bmatrix} \ddot{x}_1 \\ \ddot{x}_2 \end{Bmatrix} + \begin{bmatrix} 2k & -k \\ -k & 3k \end{bmatrix} \begin{Bmatrix} x_1 \\ x_2 \end{Bmatrix} = \begin{Bmatrix} 0 \\ 0 \end{Bmatrix} \tag{4.9}$$

The stiffness matrix $\mathbf{K}$ contains off-diagonal terms which are a result of the stiffness coupling.

Consider the system of Figure 4.3 consisting of two identical masses attached to a light rigid bar supported at its ends on two identical springs. This is a 2-DOF system because it requires two variables or coordinates to define its motion. However, these can be chosen in a variety of ways and two convenient sets are shown in the Figure. They are (1) the displacements at the two ends of the bar, x_1 and x_2, and (2) the displacement at the centre of gravity, x, and the rotation θ. If the equations of motion are written down, as before, by considering the free body diagram (FBD) of the bar (see Figure 4.3) then the first set of coordinates x_1 and x_2 can lead to equations with inertial

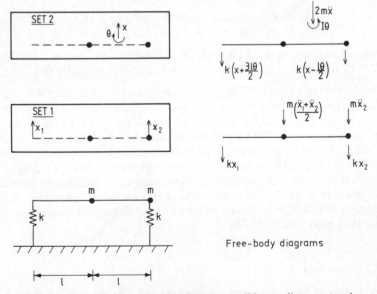

Free-body diagrams

Figure 4.3 A 2-DOF system showing two possible coordinate sets and the corresponding free body diagrams

(or mass, or dynamic) coupling because the mass matrix contains off-diagonal terms. The equation in this form is

$$\begin{bmatrix} m & m \\ m & 5m \end{bmatrix} \begin{Bmatrix} \ddot{x}_1 \\ \ddot{x}_2 \end{Bmatrix} + \begin{bmatrix} 4k & 0 \\ 0 & 4k \end{bmatrix} \begin{Bmatrix} x_1 \\ x_2 \end{Bmatrix} = \begin{Bmatrix} 0 \\ 0 \end{Bmatrix} \qquad (4.10)$$

However, dynamic equilibrium of the second FBD gives

$$\begin{bmatrix} 2m & 0 \\ 0 & I \end{bmatrix} \begin{Bmatrix} \ddot{x} \\ \ddot{\theta} \end{Bmatrix} + \begin{bmatrix} 2k & kl \\ kl & 5kl^2/2 \end{bmatrix} \begin{Bmatrix} x \\ \theta \end{Bmatrix} = \begin{Bmatrix} 0 \\ 0 \end{Bmatrix} \qquad (4.11)$$

in which there is only stiffness coupling. Of course, the actual motion is the same whichever of the two Equations (4.10) and (4.11) is used. They are different ways of expressing the same thing and can easily be shown to be equivalent. Other choices of coordinates may give both static and dynamic coupling. The question naturally arises of whether coordinates can be chosen to give no coupling at all. If this is possible, the resulting equations are each like Equation (2.1) for a simple oscillator and can be solved separately. Such coordinates can always be found and are called Principal Coordinates. The only snag is that these coordinates do not have a simple physical interpretation and may be difficult to find.

4.3 Principal coordinates

Principal coordinates for the system of Figure 4.1 will now be found. It is required to change coordinates from the directly meaningful x_1 and x_2 to a new set of coordinates y_1 and y_2 so that each of Equations (4.1) and (4.2) reduces to the simple form

$$\ddot{y} + \omega^2 y = 0 \qquad (4.12)$$

This is achieved by making y a linear combination of x_1 and x_2

$$y = x_1 + ax_2 \qquad (4.13)$$

where a is an unknown constant. Substitution of Equation (4.13) in Equation (4.12) gives

$$\ddot{x}_1 + a\ddot{x}_2 + \omega^2(x_1 + ax_2) = 0 \qquad (4.14)$$

Now add b times Equation (4.2) to Equation (4.1) and divide by m, where b is another unknown constant, giving

$$\ddot{x}_1 + 2b\ddot{x}_2 + (2 - b)\frac{k}{m}x_1 - (1 - 3b)\frac{k}{m}x_2 = 0 \qquad (4.15)$$

Comparison of the coefficients of Equations (4.14) and (4.15) will serve to define a, b and ω^2 as follows:

$$a = 2b$$
$$\omega^2 = (2 - b)k/m$$
$$\omega^2 a = (3b - 1)k/m$$

Hence by eliminating m, ω^2/k and a, a quadratic equation for b is found

$$2b^2 - b - 1 = 0$$

By simple factorization this gives two values for b and corresponding values for a, y and ω^2 as follows:

$b = +1;$ $a = +2;$ $y_1 = x_1 + 2x_2;$ $\omega_1^2 = \sqrt{k/m}$
$b = -0.5;$ $a = -1;$ $y_2 = x_1 - x_2;$ $\omega_2^2 = 1.581\sqrt{k/m}$

Hence the principal coordinates y_1 and y_2 are defined and the natural frequencies are the same as those obtained in Section 4.1. Note that mode shape information is contained within the principal coordinates. From the analysis in Section 4.1 it is known that in the fundamental mode of vibration $X_1/X_2 = 1$ and hence $y_2 = 0$; similarly, in the second mode of vibration $X_1/X_2 = -2$ and hence $y_1 = 0$. Thus, at a natural frequency of vibration only the principal coordinate corresponding to that frequency has a non-zero value.

4.4   Damping and stability

Referring to Figure 4.1 imagine that there is a damping element, with damping coefficient c, placed between mass 1 and earth and another element of damping coefficient $2c$ placed between mass 2 and earth. Equation (4.1) will then have an additional term of $c\dot{x}_1$, on the left-hand side. Equation (4.2) will similarly have an additional term of $2c\dot{x}_2$. A solution to these equations of the form

$$\begin{aligned} x_1 &= X_1 e^{\lambda t} \\ x_2 &= X_2 e^{\lambda t} \end{aligned} \tag{4.16}$$

is assumed and substituted in the equations. An analysis similar to that in Section 4.1 gives

$$\frac{X_1}{X_2} = \frac{k}{m\lambda^2 + c\lambda + 2k} = \frac{2m\lambda^2 + 2c\lambda + 3k}{k} \tag{4.17}$$

and this leads to the quartic polynomial frequency equation

$$2\alpha^4 + 8\zeta\alpha^3 + (7 + 8\zeta^2)\alpha^2 + 14\zeta\alpha + 5 = 0 \tag{4.18}$$

where $\zeta = c/2\sqrt{mk}$ and $\alpha^2 = \lambda^2 m/k$. In general, if ζ is small enough, this type of equation will have roots which are complex conjugate pairs and these will indicate, by their real parts, the damping of each mode of vibration. Also, in general (though not in this particular case), X_1/X_2 will be a complex number indicating that the motions of the two masses are not in phase. When this is so, the two modes constantly interact so that principal modes cannot be found. Equation (4.18) is a result of what is called proportional damping (see ref. 3 in Further reading, p. 115) and in this special case principal coordinates can be found. Equation (4.18) conveniently factorizes to

$$(\alpha^2 + 2\zeta\alpha + 1)(2\alpha^2 + 4\zeta\alpha + 5) = 0 \qquad (4.19)$$

If the left-hand bracket is zero a solution is found which may be interpreted as a vibration at a frequency of $\omega_n\sqrt{1 - \zeta^2}$ with a damping ratio of ζ. This constitutes the damped fundamental mode of vibration. If the right-hand bracket is zero this corresponds to the second mode of vibration with a natural frequency of $1.581\omega_n\sqrt{1 - 2\zeta^2/5}$ and a damping ratio of $\zeta/1.581$. Note that the two modes of vibration have different damping ratios and that these can be found from the complex roots of the polynomial frequency equation.

Some types of coupled motion, in which there is an external supply of energy, give rise to a polynomial equation leading to negative damping values. In such cases oscillations build up until destruction or some other limitation occurs. These cases of dynamic instability are particularly prevalent in fluid mechanics. One type is the flutter of aerofoil surfaces such as aircraft wings and turbine blades. This involves an aerodynamic coupling between bending and torsional motions of the surface. The coupling is dependent on the flow velocity and is also governed by the relative positions of the mass, structural and aerodynamic axes of the surface. An analysis of flutter is too complicated to give here but the interested reader will find more information in ref. 6, Further reading, p. 115. The famous failure of the Tacoma Narrows suspension bridge in 1940 was an unstable oscillation of a related kind.

The polynomial frequency equation can relatively easily be checked to determine whether oscillation is going to be unstable. If the equation is reduced (cf. Equation (4.18)) to

$$\alpha^4 + A\alpha^3 + B\alpha^2 + C\alpha + D = 0 \qquad (4.20)$$

then it is required to know the conditions on A, B, C and D for stability. The expression on the left-hand side of Equation (4.20) can be reduced to two quadratic factors with real coefficients (cf. Equation (4.19)) though it may prove to be difficult. Thus Equation (4.20) can be written

$$(\alpha^2 + b_1\alpha + c_1)(\alpha^2 + b_2\alpha + c_2) = 0 \tag{4.21}$$

The roots obtained from each quadratic factor are of the form

$$\alpha = (-b \pm \sqrt{b^2 - 4c})/2$$

and if the values of α so obtained are always to have negative real parts, for stability, then the conditions are

$$b_1 > 0, \qquad c_1 > 0, \qquad b_2 > 0 \qquad \text{and} \qquad c_2 > 0$$

It can be shown (see ref. 6, Further reading) that these conditions may be translated into conditions on the coefficients of Equation (4.20). These are that A, B, C and D should all be positive and that

$$(ABC - A^2D - C^2) > 0 \tag{4.22}$$

This is a specific example, for the 2-DOF system, of the more general Routh–Hurwitz stability criteria. For systems with more degrees of freedom the criteria become a great deal more complicated. Note that in determining stability it has not been necessary to find the roots of the equation.

4.5   Forced vibration: the vibration absorber

The general analysis of the harmonic forced motion of a 2-DOF system is conveniently illustrated by reference to a device, used by mechanical engineers, called a tuned vibration absorber. The vibration absorber is an auxiliary mass-spring system attached to a machine vibrating under the action of an oscillatory force, see Figure 4.4. It is designed to reduce the motion of the machine to a minimum. This is achieved by making the natural frequency of the auxiliary system (the absorber) equal to the driving frequency. Under these conditions the motion of the machine is zero, as will be shown shortly. This type of absorber, the undamped vibration absorber, is particu-

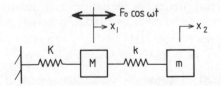

Figure 4.4 Model for a machine system with vibration absorber attached

larly suitable when the exciting frequency is of a fixed and constant value, as it would be, for example, in the case of machines driven by a.c. induction motors. Absorbers are normally used when the machine is close to resonance with the applied force. It is usual therefore in the theory to regard the natural frequency of the main system $\omega_1 = \sqrt{K/M}$ as being the same as that of the absorber $\omega_2 = \sqrt{k/m}$ though this is not essential for the absorber to function effectively. Use of the word absorber to describe the auxiliary system may imply that energy is being absorbed. This cannot be so because for this system there is no damping. In fact, the device acts as a dynamic balancer with the inertia force due to the motion of mass m acting through spring k in precise opposition to the applied force. The equations of motion are

$$M\ddot{x}_1 + (K + k)x_1 - kx_2 = F_0 \cos \omega t$$
$$m\ddot{x}_2 + kx_2 - kx_1 = 0 \tag{4.23}$$

Equations of this kind (without damping) have a solution of the form

$$x_1 = X_1 \cos \omega t$$
$$x_2 = X_2 \cos \omega t \tag{4.24}$$

and substitution in Equations (4.23) gives two simultaneous equations for X_1 and X_2 which have solutions

$$X_1 = X_2 \left[1 - \left(\frac{\omega}{\omega_2} \right)^2 \right]$$

$$X_2 = \frac{F_0/K}{\left\{ \left[1 - \left(\frac{\omega}{\omega_2} \right)^2 \right] \left[1 + \frac{k}{K} - \left(\frac{\omega}{\omega_1} \right)^2 \right] - \frac{k}{K} \right\}} \tag{4.25}$$

The motion of the main system (X_1) is always zero when the natural frequency of the absorber (ω_2) is equal to the driving frequency (ω). Under this condition the motion of the absorber is $X_2 = -F_0/k$. Because the absorber is usually small by comparison with the main machine this motion can be large and this will often be the critical factor in the choice of mass m for the absorber.

The addition of the absorber to the machine leads to two natural frequencies for the whole system. Because the absorber is relatively small, these two frequencies will be slightly above and slightly below the natural frequency of the machine on its own (see worked Example 4.2). Thus if the driving frequency ω is variable over a range close to ω_1 there is a danger of resonant motion at two driving frequencies. To avoid this problem a viscous damper with coefficient c is interposed between M and m. The equations of motion (4.23) are modified to

$$M\ddot{x}_1 + c\dot{x}_1 + (K + k)x_1 - c\dot{x}_2 - kx_2 = F_0 \cos \omega t$$
$$m\ddot{x}_2 + c\dot{x}_2 + kx_2 - c\dot{x}_1 - kx_1 = 0 \tag{4.26}$$

Equations of this kind (with damping) are most easily evaluated by means of a complex number solution, as in Chapter 3, Section 3.4, of the form

$$z_1 = X_1\, e^{i(\omega t - \phi_1)}$$
$$z_2 = X_2\, e^{i(\omega t - \phi_2)} \tag{4.27}$$

The details of the further analysis of these equations is too long to be given here but can be found in Further reading, p. 115, ref. 6, p. 93. The interesting results obtained are that

(1) The best tuning of the absorber system is at a frequency

$$\omega_2 = \omega_1/(1 + \mu) \tag{4.28}$$

where $\mu = m/M$ and is called the mass ratio. It is usually small, of the order of 0.1 to 0.01.

(2) The optimum damping ratio for the absorber, derived from the condition given by den Hartog, is given by

$$c/2\sqrt{mk} = \zeta = [3\mu/8(1 + \mu)]^{1/2} \tag{4.29}$$

There is an extensive and interesting literature in the application of vibration absorbers to engineering problems. Apart from the references to Further reading already cited the reader is encouraged to study ref. 5 which contains accounts of a number of interesting modern developments in the subject.

WORKED EXAMPLES

Example 4.1   Vibration properties of an undamped 2-DOF system

Write a program to determine the natural frequencies and mode shapes for a general undamped 2-DOF system. Use the program to solve the problem for the bending vibrations of an aircraft wing which can be simply modelled as two masses with connection springs as shown in Figure 4.5(b). The appropriate masses and stiffnesses are

$m_1 = 2500\,\text{kg}$
$m_2 = 1100\,\text{kg}$
$k_1 = 1000\,\text{kN/m}$
$k_2 = 400\,\text{kN/m}$

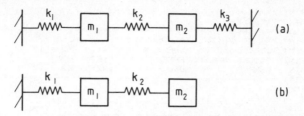

Figure 4.5 A general 2-DOF system; two typical forms

Theory

Simple 2-DOF models are usually defined in one of two ways, depending on whether there are one or two earth connections, as shown in Figure 4.5. A detailed analysis is only required for type (a) because type (b) can be regarded as type (a) with $k_3 = 0$. Analysis similar to that in Section 4.1, p. 54, gives a frequency equation

$$m_1 m_2 \omega^4 - [m_1(k_2 + k_3) + m_2(k_1 + k_2)]\omega^2$$
$$+ k_1 k_2 + k_1 k_3 + k_2 k_3 = 0 \qquad (4.30)$$

and a modal equation

$$\frac{X_1}{X_2} = 1 + \frac{k_3}{k_2} - \frac{m_2 \omega^2}{k_2} \qquad (4.31)$$

These two Equations (4.30) and (4.31) are used to find the natural frequencies and their corresponding mode shapes.

```
10   REM  EX4POINT1
20   PRINT "VIBRATION PROPERTIES OF AN UNDAMPED 2-DOF SYSTEM"
30   PRINT "-----------------------------------------------------"
40   PRINT
50   PRINT
60   PRINT "     USE CONSISTENT UNITS"
70   PRINT
80   PRINT
90   PRINT "INPUT MASS VALUES M1, M2 ?"
100  INPUT M1,M2
110  PRINT "INPUT STIFFNESS VALUES K1, K2, K3 ?"
120  INPUT K1,K2,K3
130  A = M1 * M2
140  B = M1 * (K2 + K3) + M2 * (K1 + K2)
150  C = K1 * K2 + K1 * K3 + K2 * K3
160  P = K3 / K2
170  Q = M2 / K2
180  D = B * B - 4 * A * C
190  E =  SQR (D)
200  H1 = (B - E) / (2 * A)
210  R1 = 1 + P - (Q * H1)
220  H2 = (B + E) / (2 * A)
230  R2 = 1 + P - (Q * H2)
240  F1 =  SQR (H1) / (2 * 3.14159)
250  F2 =  SQR (H2) / (2 * 3.14159)
```

```
260  PRINT  "--------"
270  PRINT  "RESULTS"
280  PRINT  "-----------------------------------------------------"
290  PRINT  "FIRST NATURAL FREQUENCY = ";F1;" HZ"
300  PRINT  "CORRESPONDING MODE : X1 = ";R1;"   X2 = 1"
310  PRINT  "SECOND NATURAL FREQUENCY = ";F2;" HZ"
320  PRINT  "CORRESPONDING MODE : X1 = ";R2;".  X2 = 1"
330  PRINT  "-----------------------------------------------------"
340  STOP

]RUN
VIBRATION PROPERTIES OF AN UNDAMPED 2-DOF SYSTEM
----------------------------------------------------

        USE CONSISTENT UNITS

INPUT MASS VALUES M1, M2 ?
?2500,1100
INPUT STIFFNESS VALUES K1, K2, K3 ?
?1E6,4E5,0
--------
RESULTS
----------------------------------------------------
FIRST NATURAL FREQUENCY = 2.25861519 HZ
CORRESPONDING MODE : X1 = .446170371   X2 = 1
SECOND NATURAL FREQUENCY = 4.277228 HZ
CORRESPONDING MODE : X1 = -.986170371   X2 = 1
----------------------------------------------------

BREAK IN 340
```

Program note

(1) The program solution to the quadratic Equation (4.30) is an abbreviated version of the program in Example 3.1. The program as a whole is a very simple one with no branches or loops. The mode shape has been normalized with $X_2 = 1$ so that Equation (4.31) can be used directly to give the value for X_1.

Example 4.2   Natural frequencies of a system with vibration absorber

It has been shown that very large motion of systems containing vibration absorbers can occur at frequencies above and below the natural frequency of the absorber, because of resonances of the system as a whole. Write a program which evaluates the system natural frequencies either (1) for specific values of machine mass, stiffness and absorber mass, or (2) in non-dimensional terms, as a ratio of system natural frequency to machine natural frequency, for a specified range of mass ratios. In both cases the natural frequencies of the machine alone and absorber alone are to be assumed identical. Two trial runs are to be made, as follows, corresponding to the two alternative program options.

(1) For a machine with $M = 220\,kg$, $K = 50\,kN/m$ and with an absorber mass $m = 22\,kg$.

(2) For a range of mass ratios from zero to 0.5 in steps of 0.05.

Theory

Using the notation of Section 4.5, p. 62, the mass ratio

$$\mu = \frac{m}{M} = \frac{k}{K}$$

and the natural frequencies are found from Equation (4.25). They are given by the condition that

$$\left[1 - \left(\frac{\omega}{\omega_2}\right)^2\right]\left[1 + \frac{k}{K} - \left(\frac{\omega}{\omega_1}\right)^2\right] - \frac{k}{K} = 0$$

but $\omega_1 = \omega_2$, and so the equation becomes

$$\left[1 - \left(\frac{\omega}{\omega_1}\right)^2\right]\left[1 + \mu - \left(\frac{\omega}{\omega_1}\right)^2\right] - \mu = 0$$

or

$$\left(\frac{\omega}{\omega_1}\right)^2 = 1 + \frac{\mu}{2} \pm \left(\mu + \frac{\mu^2}{4}\right)^{1/2} \qquad (4.32)$$

```
10   REM  EX4POINT2
20   PRINT "NATURAL FREQUENCIES OF A SYSTEM WITH VIBRATION ABSORBER"
30   PRINT "-----------------------------------------------------"
40   PRINT "WHICH CALCULATION DO YOU WANT ?"
50   PRINT "            SPECIFIC CASE ?          (1)"
60   PRINT "            RANGE OF MASS RATIOS ?   (2)
70   INPUT I
80   IF I = 1 THEN 220
90   PRINT "INPUT START, FINISH, INTERVAL FOR MASS RATIO RANGE ?"
100  INPUT P,Q,R
110  PRINT "---------------------------------"
120  PRINT "RESULTS AS FREQUENCY RATIOS (FR)"
130  PRINT "--------------------------------"
140  PRINT "MASS RATIO","LOWER FR","UPPER FR"
150  PRINT "--------------------------------"
160  FOR MR = P TO Q STEP R
170  GOSUB 380
180  PRINT MR,S,T
190  NEXT MR
200  PRINT "--------------------------------"
210  GOTO 370
220  PRINT "INPUT MAIN SYSTEM MASS, STIFFNESS, ABSORBER MASS"
230  PRINT "            * USE S.I. UNITS *"
240  INPUT M1,K1,M2
250  F1 = SQR (K1 / M1) / (2 * 3.14159)
260  MR = M2 / M1
270  GOSUB 380
```

```
280  PRINT "--------"
290  PRINT "RESULTS"
300  PRINT "---------------------------------------------------"
310  PRINT "MAIN SYSTEM NATURAL FREQUENCY = ";F1;" HZ"
320  PRINT
330  PRINT "NATURAL FREQUENCIES FOR THE SYSTEM WITH ABSORBER -"
340  PRINT "    FIRST NF = ";F1 * S;" HZ (RATIO = ";S;")"
350  PRINT "    SECOND NF = ";F1 * T;" HZ (RATIO = ";T;")"
360  PRINT "---------------------------------------------------"
370  STOP
380  A = 1 + MR / 2
390  B =  SQR (MR + MR * MR / 4)
400  S =  SQR (A - B)
410  T =  SQR (A + B)
420  RETURN


]RUN
NATURAL FREQUENCIES OF A SYSTEM WITH VIBRATION ABSORBER
-------------------------------------------------------
WHICH CALCULATION DO YOU WANT ?
         SPECIFIC CASE ?            (1)
         RANGE OF MASS RATIOS ?    (2)
?1
INPUT MAIN SYSTEM MASS, STIFFNESS, ABSORBER MASS
            * USE S.I. UNITS *
?220,50000,22
--------
RESULTS
-------------------------------------------------------
MAIN SYSTEM NATURAL FREQUENCY = 2.39935307 HZ

NATURAL FREQUENCIES FOR THE SYSTEM WITH ABSORBER -
    FIRST NF = 2.04978881 HZ (RATIO = .854308954)
    SECOND NF = 2.80853087 HZ (RATIO = 1.17053672)
-------------------------------------------------------


BREAK IN 370


]RUN
NATURAL FREQUENCIES OF A SYSTEM WITH VIBRATION ABSORBER
-------------------------------------------------------
WHICH CALCULATION DO YOU WANT ?
         SPECIFIC CASE ?            (1)
         RANGE OF MASS RATIOS ?    (2)
?2
INPUT START, FINISH, INTERVAL FOR MASS RATIO RANGE ?
?0,0.5,0.05
-----------------------------------
RESULTS AS FREQUENCY RATIOS (FR)
-----------------------------------
```

MASS RATIO	LOWER FR	UPPER FR
0	1	1
.05	.894427191	1.11803399
.1	.854308954	1.17053672
.15	.824928272	1.21222661
.2	.801088279	1.24830187
.25	.780776406	1.28077641
.3	.762960789	1.31068335
.35	.747028692	1.33863667
.4	.732581082	1.36503661
.45	.719340959	1.39016135

```
BREAK IN 370
```

Program notes

(1) A subroutine is used here which calculates the two values (S and T) for the natural frequency ratio ω/ω_1 from Equation (4.32). It is the core calculation for both program options and is therefore a suitable candidate for a subroutine even though in this case it saves only one line of program.

(2) In the first sample run the machine has a natural frequency of 2.4 Hz so an absorber is fitted which is tuned to this frequency. This suppresses all motion of the machine at 2.4 Hz. However, the calculation shows that resonant motion of the system can occur at the nearby frequencies of 2.0 Hz and 2.8 Hz. The second sample run shows how this effect depends on mass ratio. The smaller the mass ratio the closer are the two natural frequencies. This may lead to a stringent demand for accurate tuning of the absorber system. By the same token a damped absorber will be required when the forcing frequencies are variable across a band (see the worked problem in Example 4.4).

Example 4.3   Free motion of an undamped 2-DOF system

Write a general program to determine the free motion of the system in Figure 4.5(a) subsequent to given initial values of displacements and velocities. The program is to be used to find the motion of the system shown in Figure 4.1 with $m = 100\,\text{kg}$ and $k = 50\,\text{kN/m}$. The given initial conditions are $x_2 = v_1 = v_2 = 0$ and $x_1 = 10\,\text{mm}$. The time variation of x_1 and x_2 is to be printed out in tabular form.

Theory

The general solution for motion is given by Equations (4.7). The ratio of A_1/A_2 is the modal ratio R_1 for the first natural frequency and B_1/B_2 is the modal ratio R_2 for the second natural frequency. Hence Equations (4.7) can be written

$$
\begin{aligned}
x_1 &= AR_1 \cos(\omega_1 t + \phi_1) + BR_2 \cos(\omega_2 t + \phi_2) \\
x_2 &= A \cos(\omega_1 t + \phi_1) + B \cos(\omega_2 t + \phi_2)
\end{aligned}
\tag{4.33}
$$

and in these equations A and B represent the contributions from the two natural modes of vibration. If the initial conditions on displacement and velocity are known at time $t = 0$ and if they are denoted by a bar then from Equations (4.33), for displacement

$$
\begin{aligned}
\bar{x}_1 &= AR_1 \cos\phi_1 + BR_2 \cos\phi_2 \\
\bar{x}_2 &= A \cos_1 + B \cos\phi_2
\end{aligned}
\tag{4.34}
$$

and for velocity

$$\bar{v}_1 = -AR\omega_1 \sin\phi_1 - BR_2\omega_2 \sin\phi_2$$
$$\bar{v}_2 = -A\omega_1 \sin\phi_1 - B\omega_2 \sin\phi_2$$
$$(4.35)$$

From these two equations the values of A, B, ϕ_1 and ϕ_2 can be found via the intermediate step of finding $A\cos\phi_1$, $A\sin\phi_1$, $B\cos\phi_2$ and $B\sin\phi_2$. By back substitution in Equations (4.33) the motion is then determined.

```
10   REM  EX4POINT3
20   PRINT "FREE MOTION - UNDAMPED TWO DEGREE OF FREEDOM SYSTEM"
30   PRINT "--------------------------------------------------------"
40   PRINT
50   PRINT "               * USE S.I. UNITS *"
60   PRINT
70   PRINT "INPUT MASS VALUES M1, M2 ?"
80   INPUT M1,M2
90   PRINT "INPUT STIFFNESS VALUES K1, K2, K3 ?"
100  INPUT K1,K2,K3
110  PRINT "INPUT INITIAL VALUES OF X1, X2, V1, V2 ?"
120  INPUT X1,X2,V1,V2
130  A = M1 * M2
140  B = M1 * (K2 + K3) + M2 * (K1 + K2)
150  C = K1 * K2 + K1 * K3 + K2 * K3
160  P = K3 / K2
170  Q = M2 / K2
180  D = B * B - 4 * A * C
190  E =  SQR (D)
200  H1 = (B - E) / (2 * A)
210  R1 = 1 + P - (Q * H1)
220  H2 = (B + E) / (2 * A)
230  R2 = 1 + P - (Q * H2)
240  W1 =  SQR (H1)
250  W2 =  SQR (H2)
260  P = R2 - R1
270  A = (R2 * X2 - X1) / P
280  B = (X1 - R1 * X2) / P
290  C = (V1 - V2 * R2) / (W1 * P)
300  D = (R1 * V2 - V1) / (W2 * P)
310  P =  SQR (A * A + C * C)
320  Q =  SQR (B * B + D * D)
330  GOSUB 500
340  R = S
350  A = B
360  C = D
370  GOSUB 500
380  PRINT "--------"
390  PRINT "RESULTS"
400  PRINT "--------------------------------------------------------"
410  PRINT "TIME","X1","X2"
420  PRINT "--------------------------------------------------------"
430  FOR T = 0 TO 1.01 STEP .05
440  X1 = P * R1 *  COS (W1 * T + R) + Q * R2 *  COS (W2 * T + S)
450  X2 = P *  COS (W1 * T + R) + Q *  COS (W2 * T + S)
460  PRINT T,X1,X2
470  NEXT T
480  PRINT "--------------------------------------------------------"
490  STOP
500  IF A = 0 THEN 560
510  IF A > 0 THEN 540
520  S =  ATN (C / A) + 3.14159
```

```
530  GOTO 600
540  S =  ATN (C / A)
550  GOTO 600
560  IF C > O THEN 590
570  S =  - 3.14159 / 2
580  GOTO 600
590  S = 3.14159 / 2
600  RETURN
```

```
JRUN
FREE MOTION - UNDAMPED TWO DEGREE OF FREEDOM SYSTEM
----------------------------------------------------

            * USE S.I. UNITS *

INPUT MASS VALUES M1, M2 ?
?100,200
INPUT STIFFNESS VALUES K1, K2, K3 ?
?5E4,5E4,1E5
INPUT INITIAL VALUES OF X1, X2, V1, V2 ?
?0.01,0,0,0
--------
RESULTS
----------------------------------------------------
TIME          X1                 X2
----------------------------------------------------
0             9.99999999E-03     0
.05           1.53525129E-04     2.11049348E-03
.1            -8.2136061E-03     1.02043867E-03
.15           4.55759574E-04     -5.1154035E-03
.2            3.9091706E-03      -3.14432724E-03
.25           -2.99018572E-03    5.34170845E-03
.3            5.08500974E-04     4.30090477E-03
.35           6.63662502E-03     -3.17961176E-03
.4            -2.98897725E-03    -2.939317E-03
.45           -9.20976523E-03    5.87034587E-04
.5            3.2017982E-03      -6.82318432E-04
.55           8.73205158E-03     4.55490688E-04
.6            -2.54918504E-03    4.57436828E-03
.65           -4.94862991E-03    5.39780259E-04
.7            2.85293735E-03     -6.41877351E-03
.75           -3.82514601E-04    -2.24198761E-03
.800000001    -4.75736902E-03    5.24213748E-03
.850000001    4.661873E-03       2.60755035E-03
.900000001    7.10179204E-03     -2.09365489E-03
.950000001    -6.21130218E-03    -5.57891988E-04
1             -7.7634616E-03     -7.80753028E-04
----------------------------------------------------

BREAK IN 490
```

Program notes

(1) The early part of this program is taken directly from Example 4.1 and calculates the natural frequencies (W1 and W2) and mode shapes (R1 and R2) for the system. It extends as far as line 250.

(2) The subroutine at line 500 is used twice and evaluates the phase angle ϕ from values for $A \cos \phi$ and $A \sin \phi$. Because of the need to use the ATN function this becomes quite a complicated program segment.

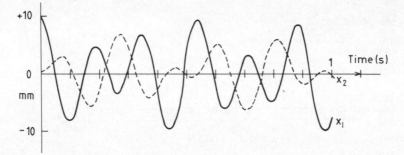

Figure 4.6 A plot of results from Example 4.3

(3) The results of the calculation are shown in Figure 4.6. It is seen that beating occurs for both motions x_1 and x_2 but the beats are out of step. Thus when motion x_1 is a maximum, x_2 is a minimum and vice-versa. This is because energy is being transferred from one mass to the other and then back again at a rate dependent on the difference between the two natural frequencies. The two natural frequencies for this system are 3.56 Hz and 5.63 Hz (use program Example 4.1 to evaluate these) and the beat rate should therefore be 2.07 Hz. Inspection of the graph for x_1 shows that this is about right. Graphics would be a useful addition to this program.

Example 4.4   Forced vibration of a damped 2-DOF system

Write a program to determine the motions of the 2-DOF system shown in Figure 4.7 subject to alternating force at point 1 only and for a range of driving frequencies. The program is to be used to study the motion of a machine system with a damped vibration absorber (see Figure 4.4, but with the addition of a damper of coefficient c fixed between M and m). The absorber has been optimally tuned and damped (see Section 4.5, p. 62) and the following values are appropriate:

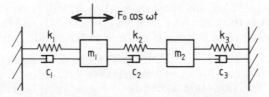

Figure 4.7 Model for the forced vibration of a damped 2-DOF system

$m_1 = M = 220\,\text{kg}$
$k_1 = K = 50\,\text{kN/m}$
$m_2 = m = 22\,\text{kg}$

The following values are calculated from Equations (4.28) and (4.29):

$k_2 = k = 4.132\,\text{kN/m}$
$c_2 = c = 111.3\,\text{Ns/m}$

For this specific problem the values shown on Figure 4.7 are:

$c_1 = c_3 = k_3 = 0$

For the purposes of calculation the force amplitude on the main machine is

$$F_0 = 100\,\text{N}$$

Theory

This type of problem is admirably suited to computer solution because the calculations are so long-winded. The theory on which the computer program is based is as follows. The equations of motion are

$$m_1\ddot{x}_1 + (c_1 + c_2)\dot{x}_1 + (k_1 + k_2)x_1 - c_2\dot{x}_2 - k_2x_2 = F_0\cos\omega t$$
$$m_2\ddot{x}_2 + (c_2 + c_3)\dot{x}_2 + (k_2 + k_3)x_2 - c_2\dot{x}_1 - k_2x_1 = 0 \qquad (4.36)$$

These have a solution of the form given in Equation (4.27) and substitution gives the following equations

$$(a + ib)A - (d + if)B = F_0$$
$$- (d + if)A + (g + ih)B = 0 \qquad (4.37)$$

where $a = k_1 + k_2 - m_1\omega^2$; $b = \omega(c_1 + c_2)$; $d = k_2$; $f = \omega c_2$; $g = k_2 + k_3 - m_2\omega^2$; $h = \omega(c_2 + c_3)$; $A = X_1 e^{-i\phi_1}$ and $B = X_2 e^{-i\phi_2}$.

In the program these two simultaneous equations are solved for A and B and the results are printed out in terms of magnitude and phases.

```
10   REM EX4POINT4
20   PRINT "FORCED MOTION - DAMPED 2-DOF SYSTEM"
30   PRINT "------------------------------------"
40   PRINT
50   PRINT "          * USE S.I. UNITS *"
60   PRINT
70   PRINT "INPUT MASS VALUES M1, M2 ?"
80   INPUT M1,M2
90   PRINT "INPUT STIFFNESS VALUES K1, K2, K3 ?"
100  INPUT K1,K2,K3
110  PRINT "INPUT DAMPING COEFFICIENT VALUES C1, C2, C3 ?"
120  INPUT C1,C2,C3
130  PRINT "INPUT FORCE AMPLITUDE ON MASS 1 ?"
```

```
140  INPUT FO
150  PRINT "INPUT FOR THE RANGE OF FREQUENCIES (IN HZ)"
160  PRINT "START, FINISH, INTERVAL ?"
170  INPUT R,S,T
180  PRINT "--------"
190  PRINT "RESULTS"
200  PRINT "-----------------------------------------------------"
210  PRINT "FREQUENCY","X1","PH 1","X2","PH 2"
220  PRINT "-----------------------------------------------------"
230  FOR F = R TO S STEP T
240  W = 2 * 3.14159 * F
250  A = K1 + K2 - (M1 * W * W)
260  B = W * (C1 + C2)
270  E = W * C2
280  G = K2 + K3 - (M2 * W * W)
290  H = W * (C2 + C3)
300  P = (A * G) - (B * H) - (K2 * K2) + (E * E)
310  Q = (B * G) + (A * H) - (2 * E * K2)
320  D = (P * P) + (Q * Q)
330  A1 = ((G * P) + (H * Q)) / D
340  A2 = ((H * P) - (G * Q)) / D
350  B1 = ((K2 * P) + (E * Q)) / D
360  B2 = ((E * P) - (K2 * Q)) / D
370  X1 = FO *  SQR ((A1 * A1) + (A2 * A2))
380  F1 =  ATN ( - A2 / A1)
390  IF F1 > 0 THEN 410
400  F1 = F1 + 3.14159
410  X2 = FO *  SQR ((B1 * B1) + (B2 * B2))
420  F2 =  ATN ( - B2 / B1)
430  IF F2 > 0 THEN 450
440  F2 = F2 + 3.14159
450  PRINT F,X1,F1,X2,F2
460  NEXT F
470  PRINT "-----------------------------------------------------"
480  STOP


]RUN
FORCED MOTION - DAMPED 2-DOF SYSTEM
------------------------------------

            * USE S.I. UNITS *

INPUT MASS VALUES M1, M2 ?
?220,22
INPUT STIFFNESS VALUES K1, K2, K3 ?
?5E4,4132,0
INPUT DAMPING COEFFICIENT VALUES C1, C2, C3 ?
?0,111.3,0
INPUT FORCE AMPLITUDE ON MASS 1 ?
?100
INPUT FOR THE RANGE OF FREQUENCIES (IN HZ)
START, FINISH, INTERVAL ?
?1.5,3,0.1
--------
RESULTS
------------------------------------------------------
```

FREQUENCY	X1	PH 1	X2	PH 2
1.50000	0.00369	0.02530	0.00651	0.22556
1.60000	0.00422	0.04781	0.00818	0.31361
1.70000	0.00501	0.09297	0.01071	0.44533
1.80000	0.00617	0.18650	0.01464	0.65321
1.90000	0.00777	0.37696	0.02031	0.99309
2.00000	0.00907	0.69577	0.02560	1.50050
2.10000	0.00888	1.01915	0.02598	2.04576
2.20000	0.00817	1.21234	0.02339	2.47324

2.30000	0.00801	1.33237	0.02123	2.81194
2.40000	0.00847	1.48034	0.01999	0.00227
2.50000	0.00909	1.72112	0.01874	0.38807
2.60000	0.00897	2.04755	0.01610	0.82541
2.70000	0.00778	2.36422	0.01220	1.22623
2.80000	0.00627	2.59680	0.00865	1.52278
2.90000	0.00501	2.74854	0.00613	1.72346
3.00000	0.00407	2.84655	0.00445	1.85914

BREAK IN 480

Program notes

(1) The results from this program are given in terms of magnitudes and phases relative to the oscillating force. The magnitude of motion for the main mass (X1) is plotted against frequency in Figure 4.8. The peak motions, which occur close to the natural frequencies found in Example 4.2, may be shown to be equal to the static deflection of M under the action of F_0 multiplied by $(1 + 2/\mu)^{1/2}$, which in this case is a factor of about 4.6. For a damped absorber there is a minimum of

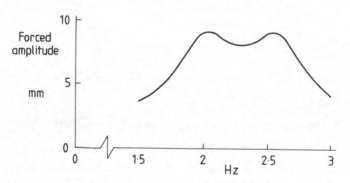

Figure 4.8 A plot of results from Example 4.4

machine motion close to its resonance frequency (2.4 Hz) rather than zero motion, as there would be for an undamped absorber.

(2) This program solves for one applied force only. It is relatively easy to extend it to solve for two applied forces, one at each mass.

(3) In this example the format of the printed results has been compressed for the same reasons as in Chapter 3.

(4) Equations (4.37) have been solved by complex number algebra specific to this problem. It would be useful, in general, to have a set of subroutines to perform simple algebraic operations on complex numbers.

(5) Note the arrangements in lines 380–400 and 420–440 inclusive to give a result for arctan (ATN) in the range 0 to π.

PROBLEMS

(4.1) Extend the analysis given in Section 4.3 to cover the case of a general 2-DOF system as shown in Figure 4.5(b). Use the algebraic results obtained to write a program which will find the principal coordinates, in terms of x_1 and x_2, together with the corresponding natural frequencies and print out the results in ascending order. Use the program to find principal coordinates for the specific vibration absorber in Example 4.2.

(4.2) A schematic two-dimensional view of a motor car is shown in Figure 4.9. The mass of the body m_1 is assumed to be concentrated at

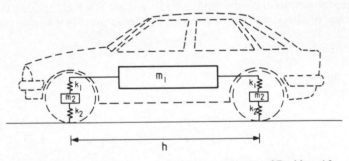

Figure 4.9 Vibration modelling for the motor car of Problem 4.2

a centre of gravity $h/3$ behind the front axle. The moment of inertia about a transverse axis is $m_1 h^2/10$. The mass of each axle is m_2. k_1 is the combined stiffness of two identical suspension springs, one on each side of the vehicle, and k_2 is the corresponding stiffness of the two inflated tyres. This system has four symmetric degrees of freedom (excluding vehicle roll) corresponding to vertical displacements on each axle, and vertical displacement and rotation (about the transverse axis) of the body. However, it can be closely modelled as a system with less degrees of freedom because $k_2 \gg k_1$. Hence, by ignoring axle motion write down the 2-DOF equations of motion which will lead to an approximate vibration analysis for heave and pitch modes. Write a program to evaluate these natural frequencies and modes of vibration. Another mode of vibration, called wheel hop, consists essentially of the axle vibrating vertically while the body remains still. This is excited primarily by the vehicle striking bumps in the road. Add an evaluation of the wheel hop frequency to the program.

The program is to be tested for a vehicle with the following properties

$m_1 = 610\,\text{kg}$
$m_2 = 72\,\text{kg}$
$k_1 = 25\,\text{kN/m}$
$k_2 = 250\,\text{kN/m}$
$h = 3.1\,\text{m}$

This problem may be analysed exactly by methods given in the next chapter. Hence, the accuracy of these simplified models can be determined.

(4.3) One of the advantages of using principal coordinates is that motion in the simple uncoupled modes can be determined very easily, the total motion being a superposition of these simple motions. Using the program obtained from Problem (4.1) as a starting section, write a program which solves the general problem given in Example 4.3. Use the program to evaluate the same trial problem and compare the results obtained with Figure 4.6. Make a comparative assessment of the two methods considering particularly the extension of the problem to three or more degrees of freedom.

(4.4) Write a program to determine the optimum properties for a damped vibration absorber given the mass and stiffness of the structure or machine to be treated and for a range of absorber masses. Make use of Equations (4.28) and (4.29).

The program is to be used to design a damped absorber for a footbridge. The absorber is to be fitted at the point of maximum vibration where the effective mass and stiffness for the first mode are 9166 kg and 1376 kN/m respectively. The excitation of the bridge by pedestrians is in the range 1.7 to 2.3 Hz and can be regarded as sinusoidal with a force amplitude of 500 N.

The relative amplitude of absorber motion should not exceed 50 mm. In order to satisfy this condition the various results for different possible mass ratios have to be substituted in a version of Example 4.4 modified to show the magnitude of relative motion $(B - A)$. This condition is often critical not only because it governs the stress in the absorber spring but also because there may be limited headroom for oscillation of the absorber mass.

(4.5) Extend Example 4.4 so that the results can be used to plot polar response diagrams for the masses m_1 and m_2 (see Example 3.2, p. 44, for a discussion of polar response diagrams). RUN the program for the sample problem in Example 4.4 but use a frequency range of 1 to 5 Hz in intervals of 0.05 Hz. Plot polar response diagrams from the results obtained (using computer graphics if possible).

(4.7) The theory in this chapter has not given any account of numerical integration of the equations of motion for transient forcing. This is because the methods are the same as those described in Section 3.6, p. 38. Extend the step-by-step integration theory of Section 3.6.1 to write a program which calculates the response of the general system in Figure 4.5(a) to a transient force applied to mass m_1 only.

In the sample RUN the vibration of an aircraft wing flying into a gust is to be determined. The wing is modelled as a 2-DOF system as in Example 4.1. The force, due to the gust, is modelled as a single half sine wave starting at $t = 0$, for which

$$F(t) = 10\,000 \sin (30t)$$

Compute and plot out the responses at both m_1 and m_2 for $0 < t < 1$. In this sample run the system has a second natural frequency of 4.3 Hz so that for reasonable accuracy and stability the integration step length should be 0.02 s or less.

Chapter 5

Systems with several degrees of freedom

The background theory for this chapter is the same as for Chapter 4 but the methods of solution are different. The relatively straightforward algebra of the previous chapter becomes more cumbersome to operate when there are more than two degrees of freedom. For systems with several degrees of freedom it is more appropriate to use matrix algebra and this is also ideally suited to computer calculations. This chapter is therefore devoted largely to matrix analysis of vibration. Some computers provide matrix instructions in BASIC though this is, as yet, uncommon on microcomputers. Some useful subroutines for performing matrix operations on arrays in BASIC have therefore been provided here.

In using computers to solve vibration problems by means of matrix methods there is one pitfall to be aware of. It is very easy to lose 'feel' for the problem when faced with large arrays of numbers on a printout of results; even more so when principal coordinates are used. The reader is therefore encouraged to make simple check calculations to ensure that computed results are of the right order of magnitude.

ESSSENTIAL THEORY

5.1 Stiffness formulation

In the previous chapter it was briefly shown in Section 4.2 that the equations of undamped motion for a 2-DOF system could be expressed in the matrix form

$$\mathbf{M\ddot{x} + Kx = 0} \tag{5.1}$$

In the usual formulation of the problem, using a mass-spring model, the mass matrix is simply a diagonal matrix containing the values for the system masses. The stiffness matrix is rather more complicated and usually contains coupling terms, as has been seen. However this matrix is easily obtained by writing down the equations of motion for a system.

If the 3-DOF system shown in Figure 5.1 is analysed in this way it is found that

$$\mathbf{K} = \begin{bmatrix} (k_1 + k_2) & -k_2 & 0 \\ -k_2 & (k_2 + k_3) & -k_3 \\ 0 & -k_3 & (k_3 + k_4) \end{bmatrix} \qquad (5.2)$$

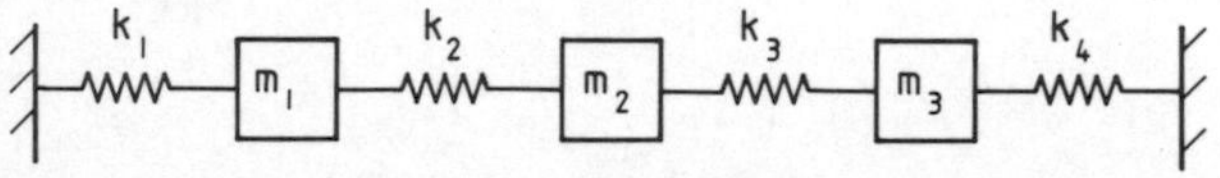

Figure 5.1 General model of an undamped 3-DOF system

This matrix is seen to have the following properties:

(1) It is symmetric, and this is an expression of Maxwell's reciprocal theorem (see Further reading, p. 115, ref. 3, p. 182).

(2) Terms on the leading diagonal are direct stiffnesses and are always positive. They are found as the sum of the stiffnesses of all springs connected to the relevant point.

(3) Terms off the diagonal are coupling stiffnesses and are always negative. They have a magnitude equal to the stiffness value for the spring connecting the two relevant points.

With these properties in mind, and after a little practice, the stiffness matrix can usually be written down by inspection.

The first step in solving Equation (5.1) is to assume a solution of the form

$$\mathbf{x} = \mathbf{X} \cos (\omega t + \phi) \qquad (5.3)$$

so that $\ddot{\mathbf{x}} = -\omega^2 \mathbf{x}$ and then

$$-\omega^2 \mathbf{M} \mathbf{X} + \mathbf{K} \mathbf{X} = 0 \qquad (5.4)$$

Premultiplying Equation (5.4) by $\mathbf{M}^{-1}$ it becomes

$$\mathbf{M}^{-1} \mathbf{K} \mathbf{X} = \omega^2 \mathbf{X} \qquad (5.5)$$

This type of equation describes the so-called matrix eigenvalue problem. The general form is

$$\mathbf{A} \mathbf{X} = \lambda \mathbf{X} \qquad (5.6)$$

where $\mathbf{A}$ is called the system matrix and λ (a scalar quantity) is called the eigenvalue. For Equation (5.5) the system matrix is $\mathbf{M}^{-1}\mathbf{K}$ and the eigenvalue is ω^2. Equation (5.6) has n distinct solutions, where n is the size of the system matrix, and each of these consists of two parts, an eigenvector $\mathbf{X}_i$ and a corresponding eigenvalue λ_i. The eigenvector $\mathbf{X}_i$

is the ith natural mode of vibration and, as might be expected, has an indeterminate scaling factor.

The determination of all the eigenvalues and eigenvectors is not an easy task. Some larger computer systems have packages which will find these values but this book assumes that there is no access to such advanced facilities. Section 5.3 will show how a limited amount of information can easily be obtained.

5.2 Flexibility formulation

This alternative formulation of the vibration problem has not previously been used in this book. It is introduced here because it has some special advantages over the stiffness approach. It requires the use of influence or flexibility coefficients which use the symbol a_{ij}. The influence coefficient a_{ij} is defined as the deflection at point i caused by the application of a unit force at point j. It is a set of descriptors alternative to the stiffness values and contains the same information in a different form. Hence there must be some connecting relationship between influence coefficients and stiffness values.

Consider the system of Figure 5.1 when each of the masses is at an extreme excursion X. In this position the inertia forces (which cause the displacements) are $m\omega^2 X$. Hence, bearing in mind the definition for the influence coefficient, equations can be written for the deflection at each point, as follows

$$X_1 = a_{11}(m_1\omega^2 X_1) + a_{12}(m_2\omega^2 X_2) + a_{13}(m_3\omega^2 X_3)$$
$$X_2 = a_{21}(m_1\omega^2 X_1) + a_{22}(m_2\omega^2 X_2) + a_{23}(m_3\omega^2 X_3) \quad (5.7)$$
$$X_3 = a_{31}(m_1\omega^2 X_1) + a_{32}(m_2\omega^2 X_2) + a_{33}(m_3\omega^2 X_3)$$

In matrix form these are written as

$$\mathbf{X} = \omega^2 \mathbf{FMX} \quad (5.8)$$

where

$$\mathbf{F} = \begin{bmatrix} a_{11} & a_{12} & a_{13} \\ a_{21} & a_{22} & a_{23} \\ a_{31} & a_{32} & a_{33} \end{bmatrix} \quad (5.9)$$

$\mathbf{F}$ is called the flexibility matrix. If Equation (5.8) is premultiplied by $\mathbf{M}^{-1}\mathbf{F}^{-1}$ then it becomes like Equation 5.5 and it follows that

$$\mathbf{K} = \mathbf{F}^{-1} \quad (5.10)$$

Thus the relationship between stiffnesses and influence coefficients has been determined.

Equation (5.8) can be rearranged to give an alternative way of writing the vibration equations which is also in the standard form of Equation (5.6)

$$\mathbf{FMX} = \frac{1}{\omega^2}\mathbf{X} \qquad (5.11)$$

Here, the system matrix $\mathbf{A}$ is given by $\mathbf{FM}$ (the inverse of the system matrix found in the last section) and the eigenvalue λ is now $1/\omega^2$ (the reciprocal of the value appropriate to the stiffness formulation). The eigenvectors remain unchanged and still constitute the mode shapes.

The first, but not obvious, advantage for the flexibility approach is that influence coefficients can be measured on a real structure. It is simply a matter of applying a known force to the structure, at each point in turn, and measuring all the relevant displacements. Stiffness values may not be measured directly. In practice these measurements can be made either on the real structure or on a properly scaled model. Another important advantage will become clear in the next section.

5.3   Simple solutions to the eigenvalue problem

In principle values for natural frequency may be found directly from Equation (5.6) because this is a set of homogeneous equations

$$(\mathbf{A} - \lambda\mathbf{I})\mathbf{X} = \mathbf{0} \qquad (5.12)$$

The condition for a solution is that the determinant of coefficients should be zero, thus

$$|\mathbf{A} - \lambda\mathbf{I}| = 0 \qquad (5.13)$$

This is called the characteristic or frequency equation. It is an nth-order polynomial equation either for ω^2 or $1/\omega^2$ depending on the formulation. As has been seen in Chapter 4, such equations can be difficult to solve. In this case it is made simpler by the knowledge that all the roots must be real and so a simple numerical technique, such as the bisection method or the secant method, can be used. See Further reading, p. 115, ref. 4, Chapter 4, for details of both these techniques and others. If the natural frequencies can be found then the corresponding mode shapes may be determined by back substitution. However, this approach is not particularly well suited to computer calculation of mode shapes.

Another technique for finding both natural frequencies and mode shapes together is the Iteration Method (sometimes known as the Power Method). Starting with a guessed vector $\mathbf{X_{g0}}$, Equation (5.6) is used to calculate $\mathbf{X_{g1}}$ as follows

$$\mathbf{X}_{g1} = \frac{1}{\lambda_{g0}} \mathbf{A} \mathbf{X}_{g0} \qquad (5.14)$$

λ_{g0} is, at this stage, a convenient normalizing factor. For example, it can be chosen to make the largest element in $\mathbf{X}_{g1}$ equal to unity. The process is repeated to find $\mathbf{X}_{g2}$ and then repeated again as many times as is necessary to achieve satisfactory convergence. A test for convergence can be made on the normalizing factors to halt the calculation when

$$|\lambda_{gn} - \lambda_{g(n-1)}| < t \qquad (5.15)$$

(but note that this test will not necessarily lead to a final value of λ accurate to within $\pm t$). The process converges to the eigenvalue of highest magnitude and to its corresponding eigenvector. In vibration terms this means that use of the method will lead to a solution for the highest value of ω^2 if the stiffness formulation is used (Equation (5.5)) and a lowest value of ω^2 if the flexibility formulation is used (Equation (5.11)). In both cases the mode of vibration will also be determined. Engineers are generally more interested in the fundamental mode of vibration than in others. This then is the second advantage of the flexibility approach.

The major drawback of the iterative method is that it finds only one of the n solutions to the n-DOF vibration problem. However this difficulty can be overcome in a number of ways. The first is to ensure that the highest order mode is always absent from the vectors $\mathbf{X}_g$. The iteration then leads to the second highest mode. This can be achieved by altering the system matrix and is a technique known as modal elimination, modal rinsing or modal sweeping. A brief account of it is given in ref. 3, Further reading (p. 115) and it also appears in refs. 8, 9 and 10. It is not a method well suited to automatic computation using BASIC.

Another method of finding all solutions is based on eigenvalue shifting (see refs. 8 and 11 for details). Use of the iteration method on Equation (5.6) will yield results for $\mathbf{X}_n$ and λ_n. The first step in finding further solutions is to form a new system matrix $\bar{\mathbf{A}}$ by subtracting $p\mathbf{X}$ from each side of Equation (5.6)

$$(\mathbf{A} - p\mathbf{I})\mathbf{X} = (\lambda - p)\mathbf{X}$$

$$\bar{\mathbf{A}}\mathbf{X} = (\lambda - p)\mathbf{X} \qquad (5.16)$$

Inverting Equation (5.16)

$$\bar{\mathbf{A}}^{-1}\mathbf{X} = \frac{1}{(\lambda - p)} \mathbf{X} \qquad (5.17)$$

Considering Equation (5.16), iteration of $\bar{\mathbf{A}}$ will lead to a mode $\mathbf{X}$ for which $(\lambda - p)$ has the greatest magnitude and this can be used, by suitable choice of p, to find the smallest value for λ. Mode determination is unaffected. More usefully, the inverse form, Equation (5.17), can be used to find all the modes. Suppose that an approximation to the ith eigenvalue, λ_{ia}, is known. Setting $p = \lambda_{ia}$ makes the magnitude of $1/(\lambda_i - p)$ large and convergence will then be to the ith eigenvalue. One difficulty with this approach is that if λ_{ia} is a good approximation $\bar{A}$ will be nearly singular and there may then be difficulty in inversion. The method also requires approximate values for all the eigenvalues, but these can be found relatively easily from Equation (5.13).

5.4 Modal orthogonality

The natural modes of vibration, given by the vectors X_i, are mutually related in a way which is easily determined from Equation (5.4). This equation, for two of the natural vibrations of order r and s, is written

$$-\omega_r^2 M X_r + K X_r = 0 \tag{5.18}$$

$$-\omega_s^2 M X_s + K X_s = 0 \tag{5.19}$$

Premultiply Equation (5.18) by X_s^T and (5.19) by X_r^T, where the superscript T denotes the transposed matrix

$$-\omega_r^2 X_s^T M X_r + X_s^T K X_r = 0 \tag{5.20}$$

$$-\omega_s^2 X_r^T M X_s + X_r^T K X_s = 0 \tag{5.21}$$

Because both M and K are symmetric matrices it follows that

$$X_r^T M X_s = X_s^T M X_r \qquad \text{and} \qquad X_r^T K X_s = X_s^T K X_r$$

and so subtracting Equation (5.21) from (5.20)

$$(\omega_s^2 - \omega_r^2) X_r^T M X_s = 0 \tag{5.22}$$

but

$$(\omega_s^2 - \omega_r^2) \neq 0$$

and so

$$X_r^T M X_s = 0$$

and it follows that $\tag{5.23}$

$$X_r^T K X_s = 0$$

These are the orthogonality relationships, so-called because of an analogy with the scalar product of ordinary two-dimensional vectors. This orthogonality also leads to the use of the term normal, to describe the natural modes of vibration.

These relationships can be used to determine the link between natural modes and principal coordinates. It will be recalled that principal coordinates (y) are a linear combination (or transformation) of primary coordinates (x) designed to decouple the equations of motion, see Section 4.3, p. 59. In general, the principal coordinates are related to the primary coordinates through a transformation matrix $\mathbf{T}$ which, in Section 4.3, was

$$\mathbf{y} = \mathbf{Tx} \quad \text{with} \quad \mathbf{T} = \begin{bmatrix} 1 & 2 \\ 1 & -1 \end{bmatrix}$$

The first step is to assemble all the natural modes (vectors) into a square matrix $\mathbf{P}$ called the modal matrix. Thus

$$\mathbf{P} = [\mathbf{X}_1, \mathbf{X}_2, \mathbf{X}_3, \ldots, \mathbf{X}_n] \tag{5.24}$$

Now this is used to make a transformation of coordinates

$$\mathbf{x} = \mathbf{Py} \tag{5.25}$$

and substituting in Equation (5.1)

$$\mathbf{MP\ddot{y}} + \mathbf{KPy} = 0$$

Premultiply this equation by $\mathbf{P}^T$ to give

$$\mathbf{P}^T\mathbf{MPy} + \mathbf{P}^T\mathbf{KPy} = 0 \tag{5.26}$$

Because of the orthogonality relationships, the matrices $\mathbf{P}^T\mathbf{MP}$ and $\mathbf{P}^T\mathbf{KP}$ are diagonal, and thus the required decoupling has been achieved. Thus Equation (5.25) represents the correct transformation and the principal coordinates are found by inverting this equation to give

$$\mathbf{T} = \mathbf{P}^{-1} \tag{5.27}$$

The terms on the diagonals of $\mathbf{P}^T\mathbf{MP}$ and $\mathbf{P}^T\mathbf{KP}$ are called the generalized masses and the generalized stiffnesses respectively.

5.5 Forced motion

Forced motion of an n-DOF system can be tackled in two ways; (1) using the direct or primary coordinates, x, and (2) using the principal coordinates, y. The first method has the advantage of giving directly meaningful solutions whilst the second has the advantage of the simplicity of solution which derives from uncoupled equations. The first method does not require a determination of the natural modes of vibration whilst the second does. The use of the principal coordinate formulation is more common, particularly for systems with many degrees of freedom, and this is one of the reasons for the emphasis placed on finding the natural frequencies and modes of vibration. The character of the forcing can be harmonic, impulsive or random. This section will deal with harmonic forcing for which either method is suitable. For impulsive or random forcing the method of principal coordinates has advantages. For example, with impulsive forcing the uncoupling of the primary equations leads to simple analyses of the type given in Section 3.6, p. 38. Another important advantage is that an analysis using a limited number of modes can be made. For example, the vibration modelling of an aircraft wing may lead to thirty degrees of freedom but it may be known that for a certain design purpose only five specific modes contribute significantly to the response. The analysis would then be made for these five principal coordinates alone.

5.5.1   Harmonic forcing using primary coordinates

If the system of Figure 5.1 is subjected to a set of forces at each mass, each varying sinusoidally at the same rate ω, but at different magnitudes, Equation (5.1) will have a right-hand side, thus

$$\mathbf{M\ddot{x}} + \mathbf{Kx} = \mathbf{F}\cos\omega t \tag{5.28}$$

where $\mathbf{F} = (F_1, F_2, F_3)$, the vector of magnitudes. The steady-state response to this forcing will be of the form

$$\mathbf{x} = \mathbf{A}\cos\omega t = (a_1, a_2, a_3)\cos\omega t \tag{5.29}$$

and substitution in Equation (5.28) gives

$$\mathbf{A} = (\mathbf{K} - \omega^2\mathbf{M})^{-1}\mathbf{F} \tag{5.30}$$

Solutions to this equation for various values for the forcing frequency, ω, give the frequency response functions for the system. In the absence of damping this will have infinite values for each of a_1, a_2 and a_3 at each of the natural frequencies. Damping may be included in this formulation though, in general, the phases of motion at the three points in the system will be different unless the damping is proportional (see Section 4.4, p. 60). A complex number formulation of the problem is most appropriate when damping is included and this leads to the complication of manipulating matrices of complex numbers. Thus

$$\mathbf{M\ddot{x}} + \mathbf{C\dot{x}} + \mathbf{Kx} = \mathbf{F}\,e^{i\omega t} \tag{5.31}$$

and the response is now of the form

$$\mathbf{x} = \mathbf{A}\,e^{i\omega t} = (a_1\,e^{-i\phi_1}, a_2\,e^{-i\phi_2}, a_3\,e^{-i\phi_3})\,e^{i\omega t} \tag{5.32}$$

where ϕ values are the phase lags of the motion with respect to the force. Substitution of Equation (5.32) in (5.31) gives

$$A = (K - \omega^2 M + i\omega C)^{-1} F \qquad (5.33)$$

This equation gives amplitudes and phases of response provided that the inverse of the complex matrix $(K - \omega^2 M + i\omega C)$ can be found. This is more complicated than finding the inverse of a real matrix. If the inverse of the complex matrix $(X + iY)$ is $(R + iS)$ then

$$R = (X + YX^{-1}Y)^{-1}$$
$$S = -(Y + XY^{-1}X)^{-1} \qquad (5.34)$$

5.5.2 Harmonic forcing using principal coordinates

Equation (5.28) is converted to principal coordinates using the modal matrix transformation given at Equation (5.28), thus

$$MP\ddot{y} + KPy = F \cos \omega t \qquad (5.35)$$

Premultiplying by P^T

$$P^T MP\ddot{y} + P^T KPy = P^T F \cos \omega t \qquad (5.36)$$

This is a series of uncoupled equations for the forced motion of simple oscillators, as in Section 2.3, p. 15, each of which has a solution like Equation (2.8). The vector $P^T F$ contains the set of generalized forces for the system.

The simplicity of the above analysis can be misleading. First, the natural modes all have to be found to assemble the matrix P. Secondly, once the solution for y is obtained it has to be reconverted to x, using Equation (5.25), in order for the result to have direct meaning.

WORKED EXAMPLES

Example 5.1   Matrix multiplication

Write a subroutine starting at line 1000 to multiply two matrices AB (in that order) with dimensions $(i \times j)$ and $(j \times k)$ respectively, placing the result in C. Incorporate the subroutine into a sample program which multiplies the matrices

$$A = \begin{bmatrix} 1 & 2 \\ 2 & 0 \\ -4 & 2 \\ 1 & -3 \end{bmatrix} \qquad B = \begin{bmatrix} 3 & 1 & -3 \\ 6 & -4 & -1 \end{bmatrix}$$

$$(4 \times 2) \qquad\qquad (2 \times 3)$$

and prints out the resulting matrix C by rows.

```
10   REM  EX5POINT1
20   PRINT "MATRIX MULTIPLICATION PROGRAM"
30   PRINT "--------------------------------------------"
40   PRINT " PRINTS PRODUCT A*B"
50   PRINT " DIMENSIONS I,J,K IN DATA LINE 310"
60   PRINT " MATRIX A BY ROWS IN DATA LINE 320"
70   PRINT " MATRIX B BY ROWS IN DATA LINE 330"
80   PRINT "--------------------------------------------"
90   DIM A(20,20),B(20,20),C(20,20)
100  READ I,J,K
110  FOR M = 1 TO I
120  FOR N = 1 TO J
130  READ A(M,N)
140  NEXT N
150  NEXT M
160  FOR M = 1 TO J
170  FOR N = 1 TO K
180  READ B(M,N)
190  NEXT N
200  NEXT M
210  GOSUB 1000
220  PRINT "RESULTS"
230  PRINT "--------------------------------------------"
240  FOR M = 1 TO I
250  FOR N = 1 TO K
260  PRINT C(M,N);"  ";
270  NEXT N
280  PRINT
290  NEXT M
300  PRINT "--------------------------------------------"
310  END
320  DATA 4,2,3
330  DATA 1,2,2,0,-4,2,1,-3
340  DATA 3,1,-3,6,-4,-1
1000  FOR M = 1 TO I
1010  FOR N = 1 TO K
1020  C(M,N) = 0
1030  NEXT N
1040  NEXT M
1050  FOR M = 1 TO I
1060  FOR N = 1 TO K
1070  FOR P = 1 TO J
1080  C(M,N) = C(M,N) + A(M,P) * B(P,N)
1090  NEXT P
1100  NEXT N
1110  NEXT M
1120  RETURN


JRUN
MATRIX MULTIPLICATION PROGRAM
--------------------------------------------
 PRINTS PRODUCT A*B
 DIMENSIONS I,J,K IN DATA LINE 310
 MATRIX A BY ROWS IN DATA LINE 320
 MATRIX B BY ROWS IN DATA LINE 330
--------------------------------------------

RESULTS
--------------------------------------------
15  -7  -5
6   2   -6
0   -12  10
-15  13  0
--------------------------------------------
```

Program notes

(1) The subroutine starting at line 1000 first clears sufficient space in array (matrix) **C** and then performs the multiplication in the usual way.

(2) Data for this program is placed in DATA lines and then READ rather than the more usual interactive method of using INPUT instructions. This has the advantage, with matrix work, of allowing a careful check of data before RUNning the program.

(3) Statement 90 permits a maximum matrix dimension of 20 but this can be increased if necessary subject to the maximum array size which the installed BASIC system allows. This technique is widely used on microcomputers. Some computers allow dynamic dimensioning of arrays (that is, a statement such as DIM(I,J)) and this is better because there is then no reservation of redundant space.

Example 5.2 Matrix inversion

Write a subroutine starting at line 2000 which inverts matrix **A**, of size $(n \times n)$, into matrix **B**. Incorporate the subroutine into a sample program which will read the data given below and print out the inverse.

$$A = \begin{bmatrix} 2 & 3 & 4 \\ 1 & 1 & 1 \\ 2 & 2 & 1 \end{bmatrix}$$

Theory

The program makes use of the Gaussian Elimination technique. This relies on the fact that a series of row operations on **A**, such as the addition of row multiples to another row, is equivalent to pre-multiplication by a matrix **T**. If these operations are designed to reduce **A** to a unit matrix then **T** must be A^{-1} because

$$TA = I \tag{5.37}$$

The algorithm contained in the following program finds **T** by reducing **A** and at the same time conducting identical operations on **B** which is initially a unit matrix, **I**. Thus while **A** is being reduced to **I**, according to Equation (5.37), **B** is changed to **TI**. The method used is to reduce **A** to an upper diagonal matrix, with ones on the diagonal, and then continue to reduce it to a unit matrix. To illustrate the process detailed steps in the procedure are shown for the given matrix below. Row operations are indicated in the left-hand column. The process can also be achieved by column operations.

OPERATIONS	A	B
START	$\begin{bmatrix} 2 & 3 & 4 \\ 1 & 1 & 1 \\ 2 & 2 & 1 \end{bmatrix}$	$\begin{bmatrix} 1 & 0 & 0 \\ 0 & 1 & 0 \\ 0 & 0 & 1 \end{bmatrix}$
$R_1 = R_1/2$ $R_3 = R_3/2$	$\begin{bmatrix} 1 & 1.5 & 2 \\ 1 & 1 & 1 \\ 1 & 1 & 0.5 \end{bmatrix}$	$\begin{bmatrix} 0.5 & 0 & 0 \\ 0 & 1 & 0 \\ 0 & 0 & 0.5 \end{bmatrix}$
$R_2 = R_2 - R_1$ $R_3 = R_3 - R_1$	$\begin{bmatrix} 1 & 1.5 & 2 \\ 0 & -0.5 & -1 \\ 0 & -0.5 & -1.5 \end{bmatrix}$	$\begin{bmatrix} 0.5 & 0 & 0 \\ -0.5 & 1 & 0 \\ -0.5 & 0 & 0.5 \end{bmatrix}$
$R_2 = R_2/-0.5$ $R_3 = R_3/-0.5$	$\begin{bmatrix} 1 & 1.5 & 2 \\ 0 & 1 & 2 \\ 0 & 1 & 3 \end{bmatrix}$	$\begin{bmatrix} 0.5 & 0 & 0 \\ 1 & -2 & 0 \\ 1 & 0 & -1 \end{bmatrix}$
$R_3 = R_3 - R_2$	$\begin{bmatrix} 1 & 1.5 & 2 \\ 0 & 1 & 2 \\ 0 & 0 & 1 \end{bmatrix}$	$\begin{bmatrix} 0.5 & 0 & 0 \\ 1 & -2 & 0 \\ 0 & 2 & -1 \end{bmatrix}$

UPPER DIAGONALIZATION COMPLETE

$R_1 = R_1 - 2R_3$ $R_2 = R_2 - 2R_3$	$\begin{bmatrix} 1 & 1.5 & 0 \\ 0 & 1 & 0 \\ 0 & 0 & 1 \end{bmatrix}$	$\begin{bmatrix} 0.5 & -4 & 2 \\ 1 & -6 & 2 \\ 0 & 2 & -1 \end{bmatrix}$
$R_1 = R_1 - 1.5R_2$	$\begin{bmatrix} 1 & 0 & 0 \\ 0 & 1 & 0 \\ 0 & 0 & 1 \end{bmatrix}$	$\begin{bmatrix} -1 & 5 & -1 \\ 1 & -6 & 2 \\ 0 & 2 & -1 \end{bmatrix}$

PROCESS COMPLETE

The method will fail if a zero appears on the diagonal and this difficulty can usually be circumvented in more elaborate programs by interchanging rows or columns.

```
10   REM  EX5POINT2
20   PRINT "MATRIX INVERSION PROGRAM"
30   PRINT "----------------------------------------"
40   PRINT " INVERTS MATRIX A INTO B"
50   PRINT " PROBLEM SIZE IN DATA LINE 300"
```

```
60    PRINT " MATRIX A BY ROWS IN DATA LINE 310 ONWARDS"
70    PRINT "----------------------------------------------"
80    DIM A(20,20),B(20,20)
90    READ N
100   FOR I = 1 TO N
110   FOR J = 1 TO N
120   READ A(I,J)
130   NEXT J
140   NEXT I
150   GOSUB 2000
160   PRINT "RESULTS"
170   PRINT "----------------------------------------------"
180   FOR I = 1 TO N
190   FOR J = 1 TO N
200   PRINT B(I,J),
210   NEXT J
220   PRINT
230   NEXT I
240   PRINT "----------------------------------------------"
250   STOP
300   DATA 3
310   DATA 2,3,4
320   DATA 1,1,1
330   DATA 2,2,1
2000  FOR I = 1 TO N
2010  FOR J = 1 TO N
2020  IF J = I THEN 2050
2030  B(I,J) = 0
2040  GOTO 2060
2050  B(I,J) = 1
2060  NEXT J
2070  NEXT I
2080  FOR K = 1 TO N
2090  IF A(K,K) = 0 THEN 2370
2100  FOR I = K TO N
2110  X = A(I,K)
2120  IF X = 0 THEN 2170
2130  FOR J = 1 TO N
2140  A(I,J) = A(I,J) / X
2150  B(I,J) = B(I,J) / X
2160  NEXT J
2170  NEXT I
2180  IF (K + 1) > N THEN 2260
2190  FOR I = (K + 1) TO N
2200  IF A(I,K) = 0 THEN 2250
2210  FOR J = 1 TO N
2220  A(I,J) = A(I,J) - A(K,J)
2230  B(I,J) = B(I,J) - B(K,J)
2240  NEXT J
2250  NEXT I
2260  NEXT K
2270  FOR K = (N - 1) TO 1 STEP - 1
2280  FOR I = K TO 1 STEP - 1
2290  X = A(I,(K + 1))
2300  FOR J = 1 TO N
2310  A(I,J) = A(I,J) - X * A((K + 1),J)
2320  B(I,J) = B(I,J) - X * B((K + 1),J)
2330  NEXT J
2340  NEXT I
2350  NEXT K
2360  GOTO 2390
2370  PRINT "ERROR -- ZERO ON DIAGONAL"
2380  STOP
2390  RETURN
```

```
JRUN
MATRIX INVERSION PROGRAM
---------------------------------------------
INVERTS MATRIX A INTO B
PROBLEM SIZE IN DATA LINE 300
MATRIX A BY ROWS IN DATA LINE 310 ONWARDS
---------------------------------------------
RESULTS
---------------------------------------------
-1            5            -1

1           -6             2

0            2            -1


---------------------------------------------

BREAK IN 250
```

Program notes

(1) The subroutine at line 2000 first sets $B = I$, ending at line 2070. It then completes the upper diagonalization of A, by row operations alone, ending at line 2260. Lines 2270 to 2350 complete the inversion, as in the worked process above, also by row operations alone. Provision is made in line 2370 for an error message in the case of a zero on the diagonal.

(2) The technique used in this program will solve most practical inversion problems. However, it is not of the most refined kind. A much fuller discussion of the use of the technique is contained in Chapter 5 of ref. 8, Further reading, p. 115.

Example 5.3   Iteration method for eigenvalues

Write a program making use of the iterative method for finding the highest eigenvalue of a matrix A together with the corresponding vector (mode shape). Write the program so that each iteration can be checked by the computer operator for convergence. Use the program to solve the following problem.

A low-speed power shaft in a ship is shown in Figure 5.2 and is modelled as five finite elements having the mass values shown

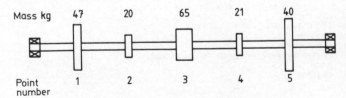

Figure 5.2 Model of the marine power shaft of Example 5.3

(derived by estimation from working drawings). The flexibility properties have been obtained by making a proper structural model, using the principles of non-dimensional analysis, and measuring the influence coefficients. The correctly scaled influence coefficients, in mm/kN, are

$$
F = \begin{bmatrix}
0.104 & 0.234 & 0.273 & 0.221 & 0.137 \\
0.234 & 0.923 & 1.333 & 1.249 & 0.735 \\
0.273 & 1.333 & 2.380 & 2.484 & 1.463 \\
0.221 & 1.249 & 2.484 & 3.212 & 2.178 \\
0.137 & 0.735 & 1.463 & 2.178 & 1.795
\end{bmatrix}
$$

Determine the fundamental whirling speed of the shaft and the corresponding mode of distortion. Note that this is a bending problem, not a torsion problem, see Section 3.7.2, p. 41.

```
10   REM  EX5POINT3
20   PRINT "ITERATION METHOD"
30   PRINT "----------------------------------------"
40   PRINT " FINDS THE HIGHEST EIGENVALUE OF A MATRIX"
50   PRINT " PROBLEM SIZE IN DATA LINE 600"
60   PRINT " FLEXIBILITY MATRIX IN DATA LINES 610-690"
70   PRINT " MASS MATRIX AS A DIAGONAL IN DATA LINE 700"
80   PRINT " ****************************************"
90   PRINT " AFTER EACH ITERATION PRESS Y TO"
100  PRINT " CONTINUE, N TO HALT EXECUTION"
110  PRINT "----------------------------------------"
120  READ N
130  DIM A(20,20),M(20),B(20),V(20)
140  FOR I = 1 TO N
150  FOR J = 1 TO N
160  READ A(I,J)
170  NEXT J
180  NEXT I
190  FOR I = 1 TO N
200  READ M(I)
210  B(I) = 1
220  FOR J = 1 TO N
230  A(J,I) = A(J,I) * M(I)
240  NEXT J
250  NEXT I
260  FOR I = 1 TO N
270  V(I) = 0
280  NEXT I
290  FOR J = 1 TO N
300  FOR I = 1 TO N
310  V(J) = V(J) + A(J,I) * B(I)
320  NEXT I
330  NEXT J
340  R = 0
350  FOR I = 1 TO N
360  IF  ABS (V(I)) < R GOTO 380
370  R =   ABS (V(I))
380  NEXT I
390  FOR I = 1 TO N
400  V(I) = V(I) / R
410  NEXT I
420  PRINT "    EIGENVALUE = ";R
```

```
430   FOR I = 1 TO N
440   B(I) = V(I)
450   NEXT I
460   GET C$
470   IF C$ = "Y" THEN 260
480   PRINT "----------------------------------------"
490   PRINT "EIGENVECTOR = ";
500   FOR I = 1 TO N
510   PRINT V(I);"  ";
520   NEXT I
530   PRINT
540   PRINT "   EIGENVALUE = ";R
550   PRINT "----------------------------------------"
560   STOP
600   DATA 5
610   DATA 1.04E-7,2.34E-7,2.73E-7,2.21E-7,1.37E-7
620   DATA 2.34E-7,9.23E-7,1.333E-6,1.249E-6,7.35E-7
630   DATA 2.73E-7,1.333E-6,2.38E-6,2.484E-6,1.463E-6
640   DATA 2.21E-7,1.249E-6,2.484E-6,3.212E-6,2.178E-6
650   DATA 1.37E-7,7.35E-7,1.463E-6,2.178E-6,1.795E-6
700   DATA 47,20,65,21,40
```

```
]RUN
ITERATION METHOD
------------------------------------------
 FINDS THE HIGHEST EIGENVALUE OF A MATRIX
 PROBLEM SIZE IN DATA LINE 600
 FLEXIBILITY MATRIX IN DATA LINES 610-690
 MASS MATRIX AS A DIAGONAL IN DATA LINE 700
 *****************************************
 AFTER EACH ITERATION PRESS Y TO
 CONTINUE, N TO HALT EXECUTION
------------------------------------------
   EIGENVALUE = 3.51399E-04
   EIGENVALUE = 2.78807258E-04
   EIGENVALUE = 2.76464938E-04
   EIGENVALUE = 2.76357085E-04
   EIGENVALUE = 2.76348569E-04
   EIGENVALUE = 2.76347662E-04
   EIGENVALUE = 2.76347558E-04
   EIGENVALUE = 2.76347545E-04
   EIGENVALUE = 2.76347544E-04
   EIGENVALUE = 2.76347544E-04
------------------------------------------
EIGENVECTOR = .094667177  .469587995  .859079028  1  .659722542
   EIGENVALUE = 2.76347544E-04
------------------------------------------

BREAK IN 560
```

Program notes

(1) For the given problem the system matrix must be found according to Equation (5.11) and this is done in lines 190 to 250. These lines can be omitted if the system matrix **A** is predetermined. In such a case lines 60 and 70 must also be changed. Note that the product **FM** is calculated in a rather unusual way taking advantage of the fact that **M** is a diagonal matrix.

(2) To start the calculation, the initial guess for the eigenvector **B** is formed in line 210 as a vector of ones.

(3) Convergence in the sample solution is rapid and at a rate of approximately one decimal place per iteration. The program can easily be altered to give automatic convergence to any required accuracy and the printout of results is then much shorter.

(4) The eigenvalue is, according to Equation (5.11), a value for $1/\omega^2$. Thus, by simple calculation, the whirl speed (natural frequency) is 574 rev/min. The mode of distortion is given by the calculated eigenvector.

(5) Note that in the DATA statements the given influence coefficients have been expressed in SI units.

Example 5.4   Harmonic forced motion of an undamped system

Write a program to determine the harmonic forced motion of an undamped n-DOF system as a function of frequency. The output of the program at each frequency is to consist of the displacement amplitudes at each point in the system. Use primary coordinates as described in Section 5.5.1, p. 86.

The program is to be used to examine the vibratory motion of a mechanism in an industrial weaving machine subject to an oscillating point force. The mechanism may be modelled as shown in Figure 5.3 and the oscillating force of magnitude 10 N is applied at point 1 on the model, as shown. The rate of oscillation of the force can vary from zero to 15 Hz.

```
10    REM EX5POINT4
20    PRINT "FORCED HARMONIC MOTION UNDAMPED SYSTEM"
30    PRINT "------------------------------------------"
40    PRINT "   PROBLEM SIZE IN DATA LINE 500"
50    PRINT "   MASS VALUES IN DATA LINE 510"
60    PRINT "   STIFFNESS MATRIX BY ROWS IN DATA LINES 520-590"
70    PRINT "   FORCE MAGNITUDES IN DATA LINE 600"
80    PRINT "NOTE: ALL FREQUENCIES ARE EXPRESSED IN RADS/S"
90    PRINT "------------------------------------------"
100   READ N
110   DIM K(20,20),A(20,20),B(20,20),M(20),X(20),F(20)
120   FOR I = 1 TO N
130   READ M(I)
140   NEXT I
150   FOR I = 1 TO N
160   FOR J = 1 TO N
170   READ K(I,J)
180   NEXT J
190   NEXT I
200   FOR I = 1 TO N
210   READ F(I)
220   NEXT I
230   PRINT "FOR FREQUENCY RANGE INPUT (START, END, INTERVAL)?"
240   INPUT R,S,T
250   PRINT "-----------------------------------------------"
260   PRINT "FREQ.","X1","X2","X3"
270   PRINT "-----------------------------------------------"
```

```
280  FOR W = R TO S STEP T
290  FOR I = 1 TO N
300  X(I) = 0
310  FOR J = 1 TO N
320   IF I = J THEN 350
330  A(I,J) = K(I,J)
340   GOTO 360
350  A(I,J) = K(I,J) - W * W * M(I)
360  NEXT J
370  NEXT I
380  GOSUB 2000
390  PRINT W,
400  FOR I = 1 TO N
410  FOR J = 1 TO N
420  X(I) = X(I) + B(I,J) * F(J)
430  NEXT J
440  PRINT X(I),
450  NEXT I
460  PRINT
470  NEXT W
480  PRINT "--------------------------------------------------"
490  STOP
500  DATA 3
510  DATA 1,1,1
520  DATA 3E3,-2E3,0,-2E3,3E3,-1E3,0,-1E3,1E3
530  DATA 10,0,0
2000  FOR I = 1 TO N
2010  FOR J = 1 TO N
2020   IF J = I THEN 2050
2030  B(I,J) = 0
2040   GOTO 2060
2050  B(I,J) = 1
2060  NEXT J
2070  NEXT I
2080  FOR K = 1 TO N
2090   IF A(K,K) = 0 THEN 2370
2100  FOR I = K TO N
2110  X = A(I,K)
2120   IF X = 0 THEN 2170
2130  FOR J = 1 TO N
2140  A(I,J) = A(I,J) / X
2150  B(I,J) = B(I,J) / X
2160  NEXT J
2170  NEXT I
2180   IF (K + 1) > N THEN 2260
2190  FOR I = (K + 1) TO N
2200   IF A(I,K) = 0 THEN 2250
2210  FOR J = 1 TO N
2220  A(I,J) = A(I,J) - A(K,J)
2230  B(I,J) = B(I,J) - B(K,J)
2240  NEXT J
2250  NEXT I
2260  NEXT K
2270  FOR K = (N - 1) TO 1 STEP - 1
2280  FOR I = K TO 1 STEP - 1
2290  X = A(I,(K + 1))
2300  FOR J = 1 TO N
2310  A(I,J) = A(I,J) - X * A((K + 1),J)
2320  B(I,J) = B(I,J) - X * B((K + 1),J)
2330  NEXT J
2340  NEXT I
2350  NEXT K
2360  GOTO 2390
2370  PRINT "ERROR -- ZERO ON DIAGONAL"
2380  STOP
2390  RETURN
```

```
]RUN
FORCED HARMONIC MOTION UNDAMPED SYSTEM
--------------------------------------------
   PROBLEM SIZE IN DATA LINE 500
   MASS VALUES IN DATA LINE 510
   STIFFNESS MATRIX BY ROWS IN DATA LINES 520-590
   FORCE MAGNITUDES IN DATA LINE 600
NOTE: ALL FREQUENCIES ARE EXPRESSED IN RADS/S
--------------------------------------------
FOR FREQUENCY RANGE INPUT (START, END, INTERVAL)?
?0,90,5
--------------------------------------------
FREQ.      X1           X2           X3
--------------------------------------------
   0    1.000E-02    1.000E-02    1.000E-02
   5    1.083E-02    1.111E-02    1.140E-02
  10    1.506E-02    1.683E-02    1.870E-02
  15    1.237E-01    1.666E-01    2.150E-01
  20   -5.932E-03   -1.271E-02   -2.118E-02
  25    6.215E-04   -4.261E-03   -1.136E-02
  30    3.836E-03   -9.713E-04   -9.713E-03
  35    8.835E-03    2.841E-03   -1.262E-02
  40    1.045E-01    6.818E-02   -1.136E-01
  45   -9.296E-03   -9.532E-03    9.299E-03
  50   -3.414E-03   -5.853E-03    3.902E-03
  55   -1.168E-03   -4.985E-03    2.461E-03
  60    5.564E-04   -5.166E-03    1.987E-03
  65    3.177E-03   -6.946E-03    2.153E-03
  70    1.873E-02   -2.280E-02    5.846E-03
  75   -1.036E-02    8.609E-03   -1.861E-03
  80   -4.638E-03    2.885E-03   -5.344E-04
  85   -3.085E-03    1.518E-03   -2.439E-04
  90   -2.329E-03    9.393E-04   -1.323E-04
--------------------------------------------

BREAK IN 490
```

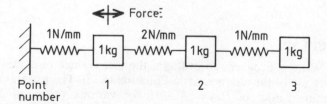

Figure 5.3 Forced motion of an undamped 3-DOF system (Example 5.4)

Program notes

(1) The program makes use of the matrix inversion subroutine devised in Example 5.2 to evaluate Equation (5.30).

(2) In the problem solved in the sample RUN the range of circular frequency W (in rad/s) is 0 to 90 and this approximates a frequency range of 0 to 15 Hz. The results are shown in Figure 5.4. The tabular printout of results is in the compressed form used previously and the heading (line 260) is specific to a 3-DOFproblem.

(3) Figure 5.4 shows that infinitely large motion occurs at the three natural frequencies of the system; W = 15.44, 40.46 and 71.59 rad/s. If realistic damping were to be included in the program (see Section 5.5.1, p. 86) then these peaks would be finite.

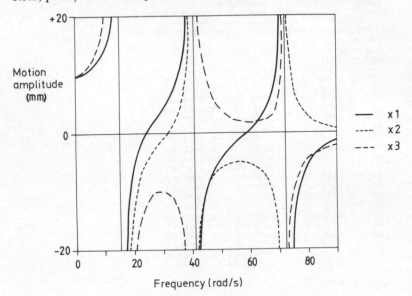

Figure 5.4 A plot of results from Example 5.4

PROBLEMS

(5.1) Write a program to evaluate the eigenvalues of a matrix **A** making use of the characteristic Equation (5.13). The program is to determine values of $f(\lambda) = |\mathbf{A} - \lambda \mathbf{I}|$ for various values of λ and converge on zeros of $f(\lambda)$ using the secant method.

Notes on techniques to be used

(1) Determinants of a matrix are most readily calculated by the Gaussian elimination technique, see Further reading, p. 115, ref. 8, p. 46 or ref. 9, p. 58. This method depends upon the value of the determinant being unaltered by simple row operations of the kind used in Example 5.2, with the exception of row factoring. If a matrix is thus reduced to upper (or lower) diagonal form the determinant is then the product of the diagonal elements. The example worked in Example 5.2 would be processed as follows

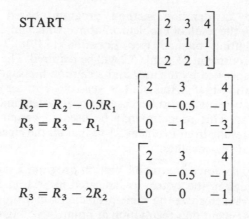

$$\text{START} \qquad \begin{bmatrix} 2 & 3 & 4 \\ 1 & 1 & 1 \\ 2 & 2 & 1 \end{bmatrix}$$

$$\begin{aligned} R_2 &= R_2 - 0.5R_1 \\ R_3 &= R_3 - R_1 \end{aligned} \qquad \begin{bmatrix} 2 & 3 & 4 \\ 0 & -0.5 & -1 \\ 0 & -1 & -3 \end{bmatrix}$$

$$R_3 = R_3 - 2R_2 \qquad \begin{bmatrix} 2 & 3 & 4 \\ 0 & -0.5 & -1 \\ 0 & 0 & -1 \end{bmatrix}$$

The determinant of the matrix is therefore $+1$. Barnett, in Further reading, ref. 9, shows how much quicker this method is than working out the determinant in the conventionally defined way. His startling illustration shows that if the determinant of an (18×18) matrix is evaluated by a computer with a multiplication time of $1\,\mu s$ the conventional calculation will take about 200 years, whereas calculation by Gaussian elimination will take 2 milliseconds. A modification to the first half of the subroutine in Example 5.2 will give a suitable program segment.

(2) The secant method for use in iteration to find zeros of $f(\lambda)$ works as follows. Two starting values λ_1 and λ_2 are chosen and $f(\lambda_1)$ and $f(\lambda_2)$ evaluated. The next best guess of a root is obtained from the equation

$$\lambda_3 = \frac{\lambda_2 f(\lambda_1) - \lambda_1 f(\lambda_2)}{f(\lambda_1) - f(\lambda_2)} \tag{5.38}$$

The process is then repeated to convergence using λ_2 and λ_3 as starting points. In the absence of other information it is not known which root has been converged upon and so it is useful to have an option in the program to print out $|\mathbf{A} - \lambda\mathbf{I}|$ for a range of values for λ. This not only indicates the order of the roots but also indicates suitable starting values for iteration to a particular root.

(5.2) Alter the iteration method program Example 5.3 to give automatic convergence to the highest eigenvalue and so eliminate the need for the options offered in lines 90 and 100. Convergence may be tested using Equation (5.15). The specified degree of convergence may be of an absolute size or, better, a specified proportion of the eigenvalue itself.

(5.3) Develop an iteration method program, based either on Example 5.3 or the result of Problem 5.2 above, which implements the eigenvalue shifting method (see Equation (5.17)). The matrix inversion subroutine in Example 5.2 will be required. The amount of shift is to be specified by the user and a caution message should be output indicating that if this shift is very close to an eigenvalue, difficulty may be experienced because the new system matrix will be nearly singular. This program may be used in conjunction with approximate results from Problem 5.1 to find all the eigenvectors as well as accurate eigenvalues.

(5.4) Develop the forced harmonic motion program Example 5.4 to include damping in the system as indicated in Section 5.5.1, p. 86. Use the program to solve the same problem as in Example 5.4 but with the requirement that the motion at point 1 (see Figures 5.3 and 5.4) shall not exceed 20 mm in amplitude in the frequency range 0 to 15 Hz. This is to be achieved by the attachment of a single viscous damper between point 1 and earth. Use the program to determine the minimum rate constant c for the damper.

(5.5) The propellor of a ship is driven directly by a diesel engine via a long propellor shaft and may vibrate excessively in torsion at some running speeds. The system is modelled as a system of discs connected by torsion springs as shown in Figure 5.5. The properties of the system are

Moments of inertia:

$$I_1 = 10 \, \text{kgm}^2, \qquad I_2 = I_3 = 25 \, \text{kgm}^2, \qquad I_4 = 200 \, \text{kgm}^2$$

Torsional stiffnesses:

$$K_{12} = 200 \, \text{Nm/rad}, \qquad K_{23} = K_{34} = 40 \, \text{Nm/rad}$$

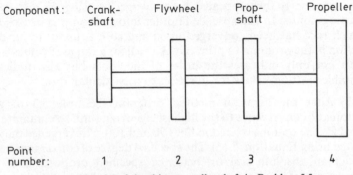

Figure 5.5 Model of the ships propeller shaft in Problem 5.5

Determine the natural frequencies and mode shapes of this system using programs from this chapter. Indicate which engine speeds in the range 0 to 800 rev/min should be avoided.

Notes on techniques to be used

(1) Torsional systems have not been analysed previously in this book. However, the equations of motion are essentially the same as those for mass-spring systems. The equation of torsional motion for a single flywheel, with moment of inertia I, attached to earth by a torsional spring of constant K is

$$I\ddot{\theta} + K\theta = 0$$

and this is exactly analogous to Equation (2.1).

(2) An additional complication arises in this problem because the system is not fixed to earth at any point. The system is nominally a 4-DOF one but, because rigid body rotation of the system is possible, there are only three degrees of freedom which involve distortion of the springs. The system is said to be degenerate and, to avoid possible numerical difficulties, the equations of motion for the system are best expressed in terms of the three relative rotations of adjacent flywheels.

Chapter 6
Bending vibrations

This final chapter is concerned with bending vibrations. Bending is perhaps the most important structural action in engineering governing the stresses, fracture and deflections in a very wide range of artefacts from the very small, such as the bimetallic strip in an electric kettle, to the very large, such as large-span box girder bridges. Dynamics is an important element in the design of beams in many cases. The vibration analysis of discrete systems, in the form of masses, springs and dampers, has been described in some detail in previous chapters. Many real structures can be modelled in this way with good accuracy, especially for the fundamental mode of vibration. This has been demonstrated by some of the worked examples and set problems earlier in the book. However, accurate modelling of all the bending modes of vibration of structure cannot be achieved by simple mass-spring representations. This is because the equations of motion are of a fundamentally different kind.

Exact solutions to the equations of motion for a structure in bending are only obtainable for beams of uniform cross-section. Most real engineering structures are not uniform and so, in design, approximate solutions are required. These are usually found in one or both of two ways. The first is to find a suitable average cross-section and then to use this in an exact analysis of the equations of motion. The second is to retain the variation of properties along the length of the beam but to use an approximate analysis based on the principle of conservation of energy. This latter technique is known as Rayleigh's Method. These two basic approaches are described here. A more advanced modern technique, which is beyond the scope of this book and the capacity of most microcomputers, is the finite element method. It is now widely used in industry to solve a whole range of problems in mechanics.

ESSENTIAL THEORY

6.1   The differential equation of motion for a uniform beam

The reader is assumed to have some basic knowledge of the theory of static bending, see Further reading, p. 115, ref. 12, Chapter 5. For the purposes of this account it is necessary to define a sign convention for the various quantities relevant to bending. This convention is defined in the sketch, Figure 6.1, of a cantilever beam subjected to a distributed downward load of w per unit length.

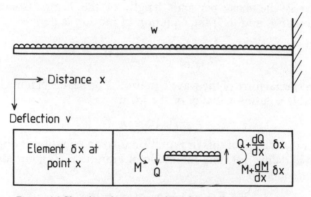

Figure 6.1 Showing sign conventions for the bending theory

The sketch of an element defines positive senses for the bending moment, M, and the shear force, Q. Equilibrium of the element gives the following relationships

$$w = \frac{\partial Q}{\partial x} \tag{6.1}$$

$$Q = \frac{\partial M}{\partial x} \tag{6.2}$$

It can also be shown that deflection, v, is related to M by the equation

$$M = EI \frac{\partial^2 v}{\partial x^2} \tag{6.3}$$

where E is the Young's modulus for the beam material and I is the second moment of area of the cross-section. The product EI is called the flexural rigidity of the beam and is a measure of its stiffness in bending. For a composite beam made up of various materials, the overall flexural rigidity is the linear sum of the EI values for each component. Combining Equations (6.1), (6.2) and (6.3)

$$w = EI \frac{\partial^4 v}{\partial x^4} \tag{6.4}$$

So far the analysis has been for static bending. Motion of the beam gives rise to inertia forces which can be regarded as a variable distributed loading, thus

$$w = -m \frac{\partial^2 v}{\partial t^2} \tag{6.5}$$

where m is the mass per unit length of the beam. Combining Equations (6.4) and (6.5) the equation of motion is derived

$$EI \frac{\partial^4 v}{\partial x^4} + m \frac{\partial^2 v}{\partial t^2} = 0 \tag{6.6}$$

This is a special form of the wave equation. For beams of finite length a separable solution is valid, of the form

$$v = f(x) \cos \omega t \tag{6.7}$$

and by substitution of this in Equation (6.6) an equation is obtained which will define not only the mode of vibration, $f(x)$, but also the natural frequency

$$EI f^{iv}(x) - m\omega^2 f(x) = 0 \tag{6.8}$$

Substituting $\lambda^4 = m\omega^2/EI$ this equation has the general solution (in two alternative forms)

$$f(\bar{x}) = A \cos \lambda x + B \sin \lambda x + C \cosh \lambda x + D \sinh \lambda x$$
$$f(x) = A e^{\lambda x} + B e^{-\lambda x} + C e^{i\lambda x} + D e^{-i\lambda x} \tag{6.9}$$

The arbitrary constants A, B, C and D do not have the same value in both forms. The first form of solution will be used hereafter. The arbitrary constants are more closely defined by application of the four end conditions, two for each end.

As an example of a solution consider the cantilever in Figure 6.1 which is assumed here to have a length l. At the left-hand end, a free end, both the shear force and the bending moment must be zero. As a consequence Equations (6.2) and (6.3) imply that at

$$x = 0; \qquad f^{iii}(x) = f^{ii}(x) = 0$$

For the right-hand (clamped) end it similarly follows that at

$$x = l; \qquad f^{i}(x) = f(x) = 0$$

(The other commonly encountered end condition is for a pinned or simply supported end. At such an end $f(x) = f^{ii}(x) = 0$.)

Substitution of these conditions in Equation (6.9) gives, in the same order

$$
\begin{aligned}
-B & & +D & &= 0 \\
-A & & +C & & &= 0 \\
-A\sin\lambda l + B\cos\lambda l &+ C\sinh\lambda l + D\cosh\lambda l &= 0 \\
A\cos\lambda l + B\sin\lambda l &+ C\cosh\lambda l + D\sinh\lambda l &= 0
\end{aligned}
\tag{6.10}
$$

This is a set of homogeneous equations. A solution is obtainable only if the determinant of the matrix of coefficients of A, B, C and D is zero. This gives the frequency equation

$$
\cos\lambda l \cosh\lambda l + 1 = 0 \tag{6.11}
$$

This equation has an infinite set of solutions each one corresponding to a natural frequency. The solutions may be found using one of the iterative methods mentioned in the last chapter, such as the secant method, to find zeros of the function

$$
f(\lambda) = \cos\lambda l \cosh\lambda l + 1 \tag{6.12}
$$

The nature of Equations (6.10) implies that no absolute solution may be found for the constants A, B, C and D but only values for their ratios. Thus, the amplitude of natural vibrations is indeterminate.

However, a normalized mode shape can be found. Equation (6.9) is rearranged thus

$$
f(x) = A\left(\cos\lambda x + \frac{B}{A}\sin\lambda x + \frac{C}{A}\cosh\lambda x + \frac{D}{A}\sinh\lambda x\right) \tag{6.13}
$$

and the term in brackets may be regarded as a normalized mode shape. The ratios are found from Equations (6.10)

$$
\frac{D}{A} = \frac{B}{A} = (\sin\lambda l - \sinh\lambda l)/(\cos\lambda l + \cosh\lambda l) \tag{6.14}
$$

$$
\frac{C}{A} = 1
$$

Hence the mode shape is determined for any particular root of Equation (6.12). This particular problem is solved numerically in worked Examples 6.1 and 6.2. Beams with other end conditions may be solved in similar fashion.

6.2 Approximate solution by Rayleigh's method

The theory developed in the previous section assumes a constant cross-section and this seldom occurs in real engineering. An alternative approach, which can easily incorporate variations in cross-section, is Rayleigh's method. This is a method based upon the principle of conservation of energy in which an approximate mode shape is chosen. It can also be used for vibrating systems other than beams in bending. In Section 2.2, p. 14 it is shown that energy is conserved in undamped vibrations. The total energy of a vibrating system ebbs and flows between kinetic energy and potential energy twice during each single cycle of vibration. The chosen mode shape is used to form integral expressions for the maximum kinetic energy and the maximum potential (strain) energy. Kinetic energy is a maximum when the beam deflections are zero, that is, when $\cos \omega t = 0$ in Equation (6.7). Differentiating Equation (6.7)

$$\dot{v} = -\omega f(x) \sin \omega t \tag{6.15}$$

$$|\dot{v}_{max}| = \omega f(x) \tag{6.16}$$

Thus the maximum kinetic energy is

$$T_{max} = \int_0^l \frac{m}{2} |\dot{v}_{max}|^2 \, dx = \frac{\omega^2}{2} \int_0^l m[f(x)]^2 \, dx \tag{6.17}$$

The maximum potential energy is stored when the beam is at a maximum excursion, with $\cos \omega t = \pm 1$ in Equation (6.7). In this condition

$$v = \pm f(x)$$

and from static bending theory the energy stored is

$$U_{max} = \frac{1}{2} \int_0^t EI[f^{ii}(x)]^2 \, dx \tag{6.18}$$

Using Equation (2.5)

$$E = U_{max} = T_{max}$$

Hence, from Equations (6.17) and (6.18)

$$\omega^2 = \frac{\int_0^l EI[f^{ii}(x)]^2 \, dx}{\int_0^l m[f(x)]^2 \, dx} \tag{6.19}$$

This equation gives an approximation to the fundamental frequency of vibration. The accuracy of the result depends upon the chosen mode $f(x)$. It is clear that the more accurate the guess for $f(x)$ the closer will the calculated frequency be to the correct value. Thus if a function is chosen which satisfies the end conditions for the beam it is likely to lead to a more accurate result. Indeed, if the correct mode is chosen the resulting frequency will be correct. The method generally leads to a high value for the natural frequency. Worked Example 6.3

gives the coefficients of suitable polynomial functions for use with Rayleigh's method.

Equation (6.19) shows that real cross-sectional variations of mass, m, and flexural rigidity, EI, will be taken into account in the integration process. In such cases analytical integration is usually impossible so that numerical integration techniques are needed. Some beamlike structures, in addition to the distributed mass m, have discrete masses, M_r, attached. An example is an aircraft wing with the mass of jet engines attached. A simple extension of Equation (6.17) shows that in such cases the frequency Equation (6.19) must have an additional term in the denominator of

$$\sum_r M_r[f(x_r)]^2 \qquad (6.20)$$

One method of improving the accuracy of the result is to use an alternative form for Equation (6.18)

$$U_{max} = \frac{1}{2} \int_0^l \frac{M_{max}^2}{EI} \, dx \qquad (6.21)$$

Here, M_{max} is obtained by twice integrating the distributed inertia forces on the beam (see Equations (6.1) and (6.2)).

$$M_{max} = \int\int m\omega^2 f(x) \, dx \qquad (6.22)$$

Note that this is an indefinite integral because M_{max} is a function of position. Methods using integrals of approximate functions generally lead to more accurate results than methods using differentials. However this improvement leads to considerably increased complexity in programming.

One of the main disadvantages of Rayleigh's method is that it only leads to a value for the fundamental frequency. This drawback can be removed by an elaboration known as the Rayleigh–Ritz method. The basis of this improvement is to choose an approximate function with n degrees of independence, thus

$$f(x) = a_1 h_1(x) + a_2 h_2(x) + a_3 h_3(x) + \cdots + a_n h_n(x) \qquad (6.23)$$

Each of the approximate functions h is chosen with the same care as before. The resulting frequency Equation (6.19) now depends on the coefficients $a_1, a_2, \ldots, a_n$. Because Rayleigh's method always gives a high value for frequency the right-hand side of Equation (6.19) can be minimized with respect to each of the coefficients a. This gives n homogeneous equations for the coefficients a. The resulting determinantal equation not only gives an improved fundamental natural frequency but also gives values for other frequencies together with good approximations to mode shapes. The method is more fully described, with an example, in ref. 3 of Further reading, p. 115.

WORKED EXAMPLES

Example 6.1   Natural frequencies of uniform beams

Write a program using Newton's iterative method to find the natural frequencies of a uniform beam.

Use the program to find the fundamental natural frequency of a cantilevered hydrofoil control surface for a submarine with the following properties

Length $= 2.2$ m
Flexural rigidity $= 2.3E7(1 - 0.1z^2) \, Nm^2$
Mass per unit length $= 500(1 - 0.1z) \, kg/m$

where z is measured from the root of the cantilever.

Theory

This example finds roots of Equation (6.12). The secant method, as described earlier, could be used to solve this problem. However in this case Newton's method is appropriate because $f(\lambda)$ is easily differentiable. The basis of Newton's method is to find both $f(\lambda)$ and the first differential $f^i(\lambda)$ at a particular guessed or starting value for λ and then to linearly project on to the λ-axis to find a closer approximation to a root, see Figure 6.2. Compare this with the secant method described in Chapter 5.

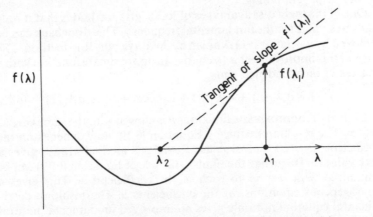

Figure 6.2 Illustrating the use of Newton's method of iteration

The improved value is

$$\lambda_2 = \lambda_1 - \frac{f(\lambda_1)}{f^i(\lambda_1)} \tag{6.24}$$

For this problem, from Equation (6.12)

$$f^i(\lambda) = l(\cos \lambda l \sinh \lambda l - \sin \lambda l \cosh \lambda l) \tag{6.25}$$

With the improved value λ_2 the iteration is repeated to convergence.

```
10   REM EX6POINT1
20   PRINT "NATURAL FREQUENCIES OF UNIFORM BEAMS"
30   PRINT "-------------------------------------------"
40   PRINT "INPUT (FLEXURAL RIGIDITY, MASS PER UNIT"
50   PRINT "LENGTH, BEAM LENGTH) ?"
60   INPUT S,M,L
70   REM  FOLLOWING FUNCTIONS DEFINED FOR A CANTILEVER BEAM
80   DEF   FN F(X) = 1 +  COS (X) * ( EXP (X) +  EXP ( - X)) / 2
90   DEF   FN FD(X) = ( COS (X) * ( EXP (X) -  EXP ( - X)) -  SIN (X) * ( EXP
     (X) +  EXP ( - X))) / 2
100  A$ = "CANTILEVER"
110   PRINT "INPUT STARTING VALUE FOR FREQUENCY (RADS/S)"
120   INPUT W
130  H = M * W * W / S
140  X = L *  SQR ( SQR (H))
150   PRINT "(CHECK ANGLE IS ";X;" RADS)"
160  Y = X -  FN F(X) /  FN FD(X)
170   IF  ABS (Y - X) < .0001 THEN 200
180  X = Y
190   GOTO 160
200  X = Y / L
210  Y = X * X *  SQR (S / M)
220   PRINT "-------------------------------------------"
230   PRINT "RESULT FOR A ";A$;" BEAM"
240   PRINT "-------------------------------------------"
250   PRINT
260   PRINT "A NATURAL FREQUENCY IS ";Y;" RADS/S"
270   PRINT
280   PRINT "-------------------------------------------"
290   PRINT "ANOTHER STARTING VALUE FOR ITERATION (Y/N) ?"
300   GET C$
310   IF C$ = "Y" THEN 110
320   STOP

]RUN
NATURAL FREQUENCIES OF UNIFORM BEAMS
--------------------------------------------
INPUT (FLEXURAL RIGIDITY, MASS PER UNIT
LENGTH, BEAM LENGTH) ?
?1.929E7,445,2.2
INPUT STARTING VALUE FOR FREQUENCY (RADS/S)
?150
(CHECK ANGLE IS 1.86734798 RADS)
--------------------------------------------
RESULT FOR A CANTILEVER BEAM
--------------------------------------------

A NATURAL FREQUENCY IS 151.248647 RADS/S

--------------------------------------------
ANOTHER STARTING VALUE FOR ITERATION (Y/N) ?

BREAK IN 320
```

Program notes

(1) Use of the DEF FN instruction allows the program to be changed, as required, to solve beam problems other than cantilevers.

(2) Line 150 helps to determine how the starting frequency stands in relation to the fundamental frequency. X is the starting value for λl. At the fundamental natural frequency λl lies between 1.5 and 5 rad/s depending on the beam end conditions. For the cantilever it is 1.875. Thus, printing of the check value helps to indicate whether the value converged upon is the fundamental natural frequency or a higher-order value. If the check value of X is much less than 1 radian convergence may be to a high order mode.

(3) The result of the calculation, 151 rad/s (or 24 Hz), is only an approximate result because of the approximations made in using average values for the stiffness and mass properties of the beam.

Example 6.2   Mode shapes for uniform beams

Write a program which will evaluate mode shapes for beam natural frequencies found in Example 6.1 and use it to find the fundamental mode of vibration for the submarine hydrofoil problem of Example 6.1.

```
10   REM EX6POINT2
20   PRINT "MODE SHAPES FOR UNIFORM BEAMS"
30   PRINT "---------------------------------------------"
40   PRINT "INPUT (FLEXURAL RIGIDITY, MASS PER UNIT"
50   PRINT "LENGTH, BEAM LENGTH) ?"
60   INPUT S,M,L
70   REM  FOLLOWING FUNCTIONS DEFINED FOR A CANTILEVER BEAM
80   DEF  FN B(X) = ( SIN (X) - ( EXP (X) -  EXP ( - X)) / 2) / ( COS (X) +
        ( EXP (X) +  EXP ( - X)) / 2)
90   A$ = "CANTILEVER"
100  DIM M(100)
110  PRINT "INPUT THE NATURAL FREQUENCY IN RADS/S ?"
120  INPUT W
130  LA =  SQR (W *  SQR (M / S))
140  PRINT "INPUT THE NUMBER OF POINTS REQUIRED ?"
150  INPUT N
160  B =  FN B(LA * L)
170  C = 1
180  D = B
190  FOR I = O TO N
200  H = LA * I * L / N
210  M(I) =  COS (H) + B *  SIN (H) + C * ( EXP (H) +  EXP ( - H)) / 2 + D
        * ( EXP (H) -  EXP ( - H)) / 2
220  NEXT I
230  X = 0
240  FOR I = O TO N
250  IF M(I) < X THEN 270
260  X = M(I)
270  NEXT I
280  FOR I = O TO N
290  M(I) = M(I) / X
300  NEXT I
310  PRINT "---------------------------------------"
320  PRINT "NORMALISED MODE SHAPE FOR A ";A$;" BEAM"
330  PRINT "POSITION","DEFLECTION"
340  PRINT "---------------------------------------"
350  FOR I = O TO N
360  PRINT (I * L / N),M(I)
370  NEXT I
380  PRINT "---------------------------------------"
390  STOP
```

```
]RUN
MODE SHAPES FOR UNIFORM BEAMS
--------------------------------------------
INPUT (FLEXURAL RIGIDITY, MASS PER UNIT
LENGTH, BEAM LENGTH) ?
?1.929E7,445,2.2
INPUT THE NATURAL FREQUENCY IN RADS/S ?
?151.248647
INPUT THE NUMBER OF POINTS REQUIRED ?
?10
--------------------------------------------
NORMALISED MODE SHAPE FOR A CANTILEVER BEAM
POSITION          DEFLECTION
--------------------------------------------
0                 1
.22               .862399544
.44               .725477692
.66               .590876298
.88               .461134554
1.1               .339523114
1.32              .229884376
1.54              .136482938
1.76              .0638709324
1.98              .0167735007
2.2               4.65661287E-10
--------------------------------------------

BREAK IN 390
```

Program notes

(1) This program makes specific use of the Equations (6.14) for a cantilever to solve this problem. This program could be appended to Example 6.1 to make a program which would find natural frequencies and corresponding mode shapes together.

(2) The loop consisting of lines 190–220 calculates modal deflections using Equation (6.13). The values of the constants B, C and D are set prior to this for the cantilever. The subsequent loops, lines 240–300, normalize the mode shape to make the greatest deflection equal to unity.

(3) The very small result for deflection at the right-hand end of the cantilever (the root, where deflection is actually zero) indicates the error level in the calculations.

Example 6.3   Polynomials for Rayleigh's method

Write a program which provides a polynomial function for use with Rayleigh's method. The function is to be the minimum necessary to satisfy all the four end conditions for any combination of simply supported, clamped or free ends. The function is to be expressed in terms of X which is the non-dimensional variable x/l.

```
10   REM EX6POINT3
20   PRINT "APPROXIMATE POLYNOMIAL FUNCTIONS FOR THE"
30   PRINT "BENDING VIBRATIONS OF BEAMS FOR USE WITH"
40   PRINT "RAYLEIGH'S METHOD"
50   PRINT "----------------------------------------"
```

```
60   DIM A$(9,2)
70   FOR I = 1 TO 9
80   FOR J = 1 TO 2
90   READ A$(I,J)
100  NEXT J
110  NEXT I
120  X = 0
130  PRINT "AT THE LEFT HAND END (X=0) IS THE BEAM"
140  PRINT "        SIMPLY-SUPPORTED ? (1)"
150  PRINT "          CLAMPED ?        (2)"
160  PRINT "      OR FREE ?            (3)"
170  INPUT N
180  IF X = 1 THEN 240
190  M = N
200  X = X + 1
210  PRINT
220  PRINT "AT THE RIGHT HAND END (X=1) IS THE BEAM"
230  GOTO 140
240  H = (3 * M - 3) + N
250  PRINT "-----------------------------------------"
260  PRINT "A POLYNOMIAL SATISFYING ALL BOUNDARY"
270  PRINT "CONDITIONS IS"
280  PRINT
290  PRINT "           ";A$(H,1)
300  PRINT
310  PRINT "THE SECOND DIFFERENTIAL FUNCTION IS"
320  PRINT
330  PRINT "           ";A$(H,2)
340  PRINT
350  PRINT "-----------------------------------------"
360  STOP
370  DATA "X^4 - 2X^3 + X","12X^2 - 12X"
380  DATA "X^4 - 1.5X^3 + 0.5X","12X^2 - 9X"
390  DATA "X^5 - 3.333X^4 + 3.333X^3","20X^3 - 40X^2 + 20X"
400  DATA "X^4 - 2.5X^3 + 1.5X^2","12X^2 - 15X + 3"
410  DATA "X^4 - 2X^3 + X^2","12X^2 - 12X + 2"
420  DATA "X^4 - 4X^3 + 6X^2","12X^2 - 24X + 12"
430  DATA "X^5 - 1.667X^4","20X^3 - 20X^2"
440  DATA "X^4 - 4X + 3","12X^2"
450  DATA "X^6 - 3X^5 + 2.5X^4","30X^4 - 60X^3 + 30X^2"

]RUN
APPROXIMATE POLYNOMIAL FUNCTIONS FOR THE
BENDING VIBRATIONS OF BEAMS FOR USE WITH
RAYLEIGH'S METHOD
-----------------------------------------
AT THE LEFT HAND END (X=0) IS THE BEAM
        SIMPLY-SUPPORTED ? (1)
          CLAMPED ?        (2)
      OR FREE ?            (3)
?3

AT THE RIGHT HAND END (X=1) IS THE BEAM
        SIMPLY-SUPPORTED ? (1)
          CLAMPED ?        (2)
      OR FREE ?            (3)
?2
-----------------------------------------
A POLYNOMIAL SATISFYING ALL BOUNDARY
CONDITIONS IS

        X^4 - 4X + 3

THE SECOND DIFFERENTIAL FUNCTION IS

        12X^2

-----------------------------------------

BREAK IN 360
```

Program notes

(1) This program primarily handles the text strings contained in the DATA statements. The algebraic hard work of evaluating the functions has been done by the programmer. Note that READ and DATA statements and arrays can all handle strings.

(2) The given deflection functions have arbitrary constant multipliers which are here chosen to make the coefficient of the highest order term equal to unity.

PROBLEMS

(6.1) Amend Example 6.1 so that roots of the frequency equation are found by the secant method of iteration, see Problem 5.1, note (2), p. 99 and compare results.

(6.2) Rework the theory in Section 6.1 for a beam which is simply supported at the left-hand end ($x = 0$) and clamped at the other end. Hence show that the appropriate frequency equation, corresponding to Equation (6.11), is

$$\cos \lambda l \sinh \lambda l - \sin \lambda l \cosh \lambda l = 0$$

Show also that $A = C = 0$ and that

$$\frac{D}{B} = -\cos \lambda l / \cosh \lambda l$$

Use these two equations to amend Examples 6.1 and 6.2 for this type of beam. Use the program to show that the fundamental natural frequency of such a beam, having otherwise the same characteristics as that used in Example 6.1, is 663.3 rad/s.

(6.3) Write a program to evaluate the fundamental natural frequency of beams using Rayleigh's method (Equation (6.19)). The program is to use a numerical integration technique, such as the trapezium rule or Simpson's rule (see Further reading, p. 115, ref. 4, p. 122ff) to evaluate the two integrals involved and it must cater for the possibility of variable mass distribution and flexural rigidity.

Use the program to find the fundamental natural frequency of the specific structure defined in Example 6.1 taking the variation of mass and flexural rigidity into account.

Notes on techniques to be used

(1) There are two sources of inaccuracy in the result obtained by this method: (a) the choice of approximate deflection function, and (b) the

integration step length. The effect of (b) can easily be gauged by computing results for different step lengths. As far as (a) is concerned, a suitable choice of deflection function is given by Example 6.3.

(2) It is good practice to use the DEF FN statement for each of the following functions: (a) mass distribution, (b) flexural rigidity, (c) chosen deflection function, and (d) the second differential function. In this way the program may easily be adapted for other beams.

Further reading

1. Leventhal, L. A., *Introduction to Microprocessors*, Prentice-Hall International, (1978).
2. Monro, D. M., *Basic BASIC*, Edward Arnold, (1978).
3. Thomson, W. T., *Theory of Vibration with Applications*, 2nd edition, George Allen & Unwin, (1981).
4. Mason, J. C., *BASIC Numerical Mathematics*, Butterworth, (1983).
5. Clough, R. W. and Penzien, J., *Dynamics of Structures*, International Student Edition, McGraw-Hill, (1982).
6. Den Hartog, J. P., *Mechanical Vibrations*, 4th edition, McGraw-Hill, (1956).
7. Hunt, J. B., *Dynamic Vibration Absorbers*, Mechanical Engineering Publications, (1979).
8. Mason, J. C., *BASIC Matrix Methods*, Butterworth, (1984).
9. Barnett, S., *Matrix Methods for Engineers and Scientists*, McGraw-Hill, (1979).
10. Seto, W. W., *Mechanical Vibrations*, McGraw-Hill Schaum Outline, (1964).
11. Jennings, A., *Matrix Computation for Engineers and Scientists*, Wiley, (1977).
12. Iremonger, M. J., *BASIC Stress Analysis*, Butterworth, (1982).

Index